Strategic Lean Mapping

Strategic Lean Mapping

Blending Improvement Processes for the Perfect Solution

Steven Borris

New York   Chicago   San Francisco
Lisbon   London   Madrid   Mexico City
Milan   New Delhi   San Juan
Seoul   Singapore   Sydney   Toronto

The McGraw·Hill Companies

McGraw-Hill books are available at special quantity discounts to use as premiums and sales promotions, or for use in corporate training programs. To contact a representative please e-mail us at bulksales@mcgraw-hill.com.

Strategic Lean Mapping: Blending Improvement Processes for the Perfect Solution

Copyright © 2012 by The McGraw-Hill Companies, Inc. All rights reserved. Printed in the United States of America. Except as permitted under the United States Copyright Act of 1976, no part of this publication may be reproduced or distributed in any form or by any means, or stored in a database or retrieval system, without the prior written permission of the publisher.

1 2 3 4 5 6 7 8 9 0   QFR/QFR   1 8 7 6 5 4 3 2

ISBN   978-0-07-178855-7
MHID   0-07-178855-7

The pages within this book were printed on acid-free paper.

Sponsoring Editor
Judy Bass

Editorial Supervisor
Stephen M. Smith

Production Supervisor
Pamela A. Pelton

Acquisitions Coordinator
Bridget L. Thoreson

Project Manager
Patricia Wallenburg, TypeWriting

Copy Editor
James Madru

Proofreader
Paul Tyler

Indexer
Claire Splan

Art Director, Cover
Jeff Weeks

Composition
TypeWriting

Information contained in this work has been obtained by The McGraw-Hill Companies, Inc. ("McGraw-Hill") from sources believed to be reliable. However, neither McGraw-Hill nor its authors guarantee the accuracy or completeness of any information published herein, and neither McGraw-Hill nor its authors shall be responsible for any errors, omissions, or damages arising out of use of this information. This work is published with the understanding that McGraw-Hill and its authors are supplying information but are not attempting to render engineering or other professional services. If such services are required, the assistance of an appropriate professional should be sought.

Writing a book ties the author to a computer for a long time and, when the writing gets tough, it is helpful to get a bit of encouragement and a glass of wine.

With that in mind, I would like to thank my wife, Carol, as she is the one who has had to put up with this—twice so far and probably a third time!

For encouragement, my nephew, David Allan, and his wife, Donna, have also been a source of constant support. Besides, their two children, Tagen and Shea, constantly tell everyone that I am a great author. This is not easy to live up to.

About the Author

Steven Borris is a manufacturing advisor and continuous improvement specialist in the science, food and drink, electronics, textile, optics, electromechanics, medicine, and semiconductor industries. His technical expertise includes equipment installation and maintenance, root-cause fault resolution, equipment redesign, calibration, customer support, commissioning, and training. Mr. Borris is currently with SMAS, a Scottish government agency tasked with improving the efficiencies of companies. He also is the author of *Total Productive Maintenance* (McGraw-Hill, 2006).

CONTENTS

Introduction.................................. xiii
Acknowledgments xxi

CHAPTER 1   **The Storm Before the Calm** 1
    Profit or Turnover? 1
    Adapting to New Techniques 2
    Diagnosis 3
    Resistance to First Projects 4
    Where Will We Get the Time to Do Projects? 9
    Productivity Tip 11
    Overall Equipment Efficiency 12
    Culture 14
    Why Is It So Important That the Initiative
        Works and Is Sustained? 16

CHAPTER 2   **Finding Improvement Opportunities** 19
    Adding *Some* Value? 20
    Right First Time 22
    Example: Root Cause 23
    Size Isn't Everything 24
    Customer Audits 25
    Where Do We Start? 25
    The Five Principles of Lean 26
    How Do We Do It? 28
    Productivity Tip: Brainstorming 30
    Mapping: General Guidance 32
    Plan-Do-Check-Act 35
    How Do I Analyze the Process? 37
    An Introduction to Process Maps 40

                Quantifying Losses 43
                The "7 Wastes"—With the Cake Icing Examples 45

CHAPTER 3   **The Big Picture Map** 49
                The SWOT Diagram 50
                The Big Picture Map versus the Process Map 52
                Production Logs 53
                Employee Input 55
                The Big Picture Map 55
                What Do We Need to Know to Make a Useful Map? . . . 57
                The Parking Lot 59
                Material Flow 61
                Kanban 63
                Materials—Step 1 64
                Information Flow—Step 2 67
                Data Flow and Data Boxes—Step 3 70
                What Is Availability? 72
                On Time and In Full Deliveries (OTIF) 76
                Lead Times—Step 4 77
                Inventory and Stocking Points—Step 5 78
                Quality Checkpoints—Step 6 78
                What Do We Need to Know? 80
                Rework and Scrap Loops—Step 7 81
                What Are the Main Issues at Each Stage? 81
                How Much Are the Problems Costing? 82
                The Future State Map 83
                Solutions 85

CHAPTER 4   **Process Mapping** 87
                Bottlenecks 88
                Mapping a Restaurant Process 89
                The Team 90
                Creating the Map 92
                Consider the Impact of the "7 Wastes" on
                    the First Eight Steps as Defined Earlier 96
                In Summary: Making the Map 104
                The Capacity Map 105

CHAPTER 5	**Capacity Mapping**........................ 107
	The Unloading Process..................... 107
	The Birth of the Capacity Map 108
	The Capacity Map......................... 110
	Improving the Capacity 114
	The Cabinet Manufacturer 116
CHAPTER 6	**Lean Manufacturing, the Value-Stream Map, and Partial Value** 123
	A Comment on Mass Production 123
	Continuous Improvement.................. 125
	Lean Manufacturing....................... 126
	Value.................................... 127
	The Value Stream 129
	Flow 131
	Pull 133
	Perfection................................ 134
	The "7 Wastes"........................... 135
	Transporting 136
	Inventory 136
	Movement 137
	Waiting.................................. 138
	Overproduction 138
	Overprocessing 140
	Defects 141
	Conclusion............................... 142
CHAPTER 7	**Problem Solving and Decision Making** 143
	The Root Cause........................... 143
	Evaluating the Cost of the Problems......... 144
	Kaizen or Define, Measure, Analyze, Improve, and Control (DMAIC) as a Problem-Solving Tool..... 147
	Step 1: Identify the Project or Problem to Be Solved............................ 147
	Step 2: Create the Improvement Team......... 149
	Step 3: Define the Problem Exactly 150

Step 4: Quantify the Problem 150
Step 5: Plan the Project 154
Step 6: Solve the Problem 155
Brainstorming 156
The "5 Whys" 157
    Example 1 157
Fishbone Diagram 159
CEDAC Diagram 159
Fault-Tree Analysis 164
    Step 7: Circulate the Results 166
    Step 8: Embed the Solution 167
What Are the Consequences of a Wrong Solution
to a Problem Being Applied? 167

Glossary 173

Index 181

INTRODUCTION

Since writing my first book, *Total Productive Maintenance*, in 2003–2004, I have had multiple opportunities to reapply the techniques covered in that book. I also have been fortunate enough to learn some new ones. *Mapping* is the key technique that I will pass on to you in this book. It is one of the most powerful ways I know to diagnose and anticipate problems.

Why Mapping?

What is *mapping*? It is a range of techniques for analyzing the steps (or stages) in a process. Any process is suitable. In my experience, the most common maps are on processes—the way a product is manufactured—but mapping also can be the way players are selected for the school football team or how a cable box is connected to a TV. Mapping is incredibly versatile. In business, we can analyze a sales process, an order process, a customer complaint process, a disciplinary process used by management, a stores process for running a warehouse, or how the maintenance or cleaning department functions; in medicine, we can analyze how a patient is prepared for surgery or how drugs are prepared and distributed to patients; and in government, we can analyze how traffic flows through an intersection or how a police officer interacts with a suspect or a victim—in short, we can map anything! Mapping even can be used to create an overview of how a company operates. I am not the first to realize this: the *big picture map*, developed by Toyota when I was a child, is one of the most useful functions around—from a diagnostic perspective. It was used originally to develop strategy at a management level, but as I will explain in Chapter 3, it can be adapted to suit whatever *you* need it to show.

Mapping is used most often to analyze a process to find ways to improve how the task is carried out. If Lean principles are applied, we can make any process more efficient. If we review the process with a team of skilled operators and engineers, we can identify any steps where errors are introduced and where they affect the rest of the process. Once we know this, we can take steps to eliminate the problems and make the process more reliable. It all boils down to what you need from the process and the degree of detail included in the map.

With some expert advice from McGraw-Hill, I chose to title this book *Strategic Lean Mapping: Blending Improvement Processes for the Perfect Solution*. I chose this title because I tend to blend key processes together in my mind. The book not only covers Lean Manufacturing and mapping, but it also introduces essential modules from the other processes I use. I have found that improvement techniques tend to have many similarities; it is usually just a few variations that make them unique. It is because of the similarities that I evolved my "pick and mix" approach. When carrying out practical problem identification and solving, I am fortunate to be able to call on a reservoir of the best bits of several techniques.

Ten years ago I had not used mapping properly, but now, as I have become more experienced in the use of mapping, I have found that I have even evolved new mapping methods based on the original methods. Why? I needed them to help identify and solve particular problems.

I am referring to my *capacity map, capacity bar chart*, and, on a different theme, my *decision flowchart*. I have even been considering a variation of the *value-stream map* (VSM). I have not used this map yet with any clients, but I have used the basic concepts to test and explain specific issues. Just as digital logic has only two states, 0 and 1, the VSM has only two states, value and no value. If value equates to a 1, no value would equate to a 0. Value is good; no value is bad. However, I don't believe that value is this simple. I believe that a task can have *some* value.

My chart introduces a third state, an analogue state, partial value. If a process step is only 50 percent efficient, to say that it adds value is true, but the amount of value added also should be considered. If only 50 percent adds value, the process also must have 50 percent that does not add value. By concentrating on only two steps, we could overlook any inefficient steps that need to be improved. It seems to be a natural progression to consider how well a given step performs related to its *capacity*—how much of the *possible* value is actually added. We are, after all, trying to improve the complete process. Partial value fits in nicely with capacity analyses, the theory of constraints, and overall equipment efficiency (OEE).

Working with the Scottish Manufacturing Advisory Service (SMAS) has brought me into contact with many more company types—bakers, brewers, tailors, carpenters, printers, builders, life science and medicine manufacturers, engineers, and cable assemblers and recyclers, to mention but a few. One fact became obvious very quickly, and it is one that I feel you must appreciate: *All* companies experience the same generic problems—no matter what the industry. Necessity is the driver that will force a company to fix any individual issues that arise, but many have never considered taking a step back and looking at the bigger picture. If they did, they could identify most of their problems, develop improved systems to prevent recurrence of similar issues, and make better profits. Productivity improvements are now promoted by government agencies, in addition to private training companies, which also teach the techniques.

Compared with 10 years ago, I now spend much more time diagnosing where systems are creating problems and how much they are costing a company. In addition, if a company progresses to an improvement project, I also need a better way to make a quick, in-depth analysis of its operation and its performance. Mapping is the perfect way to do this. When blended with components of Lean Manufacturing, reliability-centered maintenance (RCM), the 5S methodology, single-minute exchange of die (SMED), the theory of constraints (TOC), total productive maintenance (TPM), and others, it becomes even more powerful.

Introducing continuous improvement usually requires a change in the company culture. In the worst case, a negative culture can prevent a project from working. At best, it can prevent the benefits from becoming embedded. From day 1, using a mapping process will involve setting up a team of employees to find the issues that need to be addressed. This kickstarts the culture change and allows the team to use a beautifully simple analysis technique. I never believed that a $10 packet of Post-its could be of so much value.

Smaller companies have fewer resources. This gives them additional issues to consider when setting up a team to work on problems. The deployment of a team could mean lost production hours. And since making the product is what brings in the money, it is very important to make sure that *essential* production is maintained.

Best Practice?

When I worked as a full-time improvement practitioner and had my own company, I had much more time to embed techniques. With limited project time, there is more need to encourage companies to take over the responsibility for the improvements and keep the momentum going. One way to do this is to prove to companies and their employees that the processes actually work. This could mean improving quality, making people's jobs easier, reducing stress by making processes more efficient, and, of course, saving the company a tangible amount of money.

It is not essential for a company to become an expert in any particular process—although this probably will be inevitable with continued use. The goal is to develop a culture of improvement and teach the basic skills needed to help the company to apply the processes necessary to improve the way the company runs and to encourage the company to develop further.

So how does a mapping exercise work?

1. *Use the appropriate map to analyze the process as it is currently applied.* There are a few mapping techniques from which to choose.
2. *Go back over the map and highlight any problems.* Problems can be Lean waste, equipment performance, information issues, risk assessments, capacity, and so on. Normally, this step involves a summary of all the known issues.
3. *Create a list of all the problems found.*
4. *Estimate how much each problem affects the company.*
5. *Prioritize the issues and pick the vital few for immediate resolution.*

I don't think I ever placed too much demand on a company's resources, but now I give it special consideration, making sure that I stay within the company's capability. If a company has only 10 employees, it is not reasonable, for example, to take three of them off-line for three days. Break up the project goals into smaller chunks to make them more manageable. Even so, there still will be production issues, and there is still a need for training the teams in the methods needed. I have an additional consideration: Working as an external advisor and trainer, I need to work with an improvement champion who will lead the projects when I am not with the company. This requires extra coaching, training, and mentoring.

Ten years ago I wanted to make "my" companies world class. This was not an unrealistic long-term goal, but it was not always a practical first step. Everyone needs to learn how to walk before they can run. (May I apologize for using the old, folksy quotations? It took me 50 years to understand what they actually mean!) The odd thing is, I tend to be very practical. I still set an ultimate goal of perfection (e.g., zero fails), but I will settle for slower progress in the right direction. I guess that I have Scottish Enterprise (SE) to thank for my transition.

SE provides business support in all areas. If the problem diagnosed by SE involved manufacturing support and the company agreed that it was needed, a vendor would be hired to provide it. SMAS was set up in 2005. This gave SE its own specialist department capable of teaching a whole range of manufacturing techniques, including Lean methodologies. I was one of the first six practitioners. It was an SE training course that taught me the distinction between bad, good, and best practice. The course made me rethink some of the training I received from global companies. Bad practice is exactly what you would expect. Best practice is how all companies *should* work. Good practice recognizes that you are not perfect but are not making obvious mistakes.

Blending Processes

The mapping procedures in this book are simple to apply. Their objective is to visually document the process and identify any issues. However, when coupled with selected tools from Lean, TPM, and RCM, the potential savings to be found will be surprising. I will explain Lean Manufacturing and parts of other methodologies but will concentrate on the "7 wastes." Indeed, rather than have one big chapter, I will discuss them on a few occasions, each time using different examples. The key point to appreciate here is that *waste* is not just scrap; it is also time, materials, resources, and effort. Losses cause workers to take more time than they need to complete a given task, effectively filling up their working day unnecessarily.

As an illustration, I like to use the most common example I know—employees looking for tools. This is a problem that is universally accepted as being normal in a huge number of companies. I have been in companies where employees easily will spend 30 minutes a day looking for the tools they need to do their work. One company measured lost time in hours! If one employee wastes 30 minutes each day, the company is losing 2.5 hours a week. This is more than 6 percent of a 40-hour week, or three weeks in a year! Remember, this is for only one employee. To solve the problem, we need to do only two things:

1. Ensure that there are enough tools so that people have the tools they need when they need them.
2. Ensure that the tools are given a location to which they must be returned after use.

These steps, coupled with a bit of self-discipline, will eliminate the searching-time loss, and we have just won back 2.5 hours each week that now can be used to work on an improvement project or increase production. I will give other examples later.

Is Lean a Positive or Just Extra Work?

Just as I need to sell the benefits to company management, both the managers and I have to sell the process to the employees. Employees can derail the success of a project. It is not surprising, therefore, that managers need to decide carefully whether to spend their cash on a promise of something better or a bit of equipment that will do something better. Some have had very little experience in the advantages of improving processes, up-skilling, and retraining their employees.

Noninvasive interventions are not really practical, although I am still looking for a way to have a minimum-impact intervention. There always will be a need for someone to lead the improvement project. This can be an external consultant, but it still will be necessary to involve the workforce if you want the project to last when the consultant moves on. Even where there is an internal improvements champion, if the workforce is not involved, what happens if the champion leaves? It is pretty much guaranteed that unless the improvements are embedded in the daily life and culture of the organization, they will evaporate away. For prolonged success, the improvement work *never* must be regarded as extra work. The improvements need to be ongoing, like preventive maintenance programs, scheduled within a planned framework.

I am still very happy with the procedures in my first book, as testified by the readers who took the time to contact me. (Naturally, I am excluding the communications assembled using words and letters cut from newspapers.) I am also proud that my book has been used as a reference in a highly respected university course (MBA in asset management) and that I have even been quoted in final exams. I just hope that the quotes were used by the people who passed!

Company Confidentiality

In all the examples in this book, the details have been changed to conceal the identities of the companies. The losses have been scaled and the product has been changed, as has the type of company, but the main points being illustrated are both true and virtually identical. Knowing the real company and product would make no difference to the examples.

Steven Borris
steven.borris@ntlworld.com

ACKNOWLEDGMENTS

This book is based on the skills I have learned from the pioneers of continuous improvement, my personal experience, a need to solve specific customer problems, and, most importantly, techniques and advice from colleagues working in company improvement. Working with fellow improvement practitioners in SMAS was a breeding ground for good advice.

The biggest influence for this book was a SMAS colleague, Agnes Pollock, who opened my eyes to the full advantages of proper mapping as a mechanism for diagnosing issues and enabling companies to review their performance in a blame-free way.

I would also like to thank Judy Bass from McGraw-Hill and Patricia Wallenburg for their invaluable help in putting this book together.

CHAPTER 1

The Storm Before the Calm

Buying this book clearly shows that you have a desire to increase your knowledge of improvement techniques and add to your skills. Apart from an increase in confidence, new skills will make you more valuable to both your current and future employers. More value will lead to a better job and more money.

Profit or Turnover?

This book will help your company in a number of ways, too, none the least by helping it make more *profit*. Notice that the emphasis is on profit and not turnover. Turnover does not guarantee profit. Profit is the extra money you get when you subtract your running costs from what you sell. It should follow, then, that the percentage profit should be greater than the current bank interest rates; otherwise, the company would make more profit simply by depositing its money in the bank.

I will not use many formulas here. When I do, it will be to simplify a statement because a formula uses few words.

$$\text{Profit} = \text{turnover} - \text{the cost to make the product and sell it}$$

How much product a company makes and the annual turnover do not guarantee a successful company. Companies, like employees, exist only to make money. If two companies are making the same product, it will be unlikely that they are both making the same profit. The reasons for the difference are the overheads, the manufacturing process, and the operating losses.

There are seven losses that *every* company suffers to varying degrees. Lean Manufacturing defines them as the "7 wastes." I will ignore the eighth waste (untapped human potential) for the moment. Once you know what the wastes are, it is easy to look for those which affect you and then take action to minimize or eliminate them. Reducing these wastes *will* increase profit. Then there are the other advantages that follow naturally. Did I mention that in the process of increasing profit, you likely will improve the environment for employees, make better use of your factory's space, improve product quality, improve delivery performance, and improve customer satisfaction?

Who is going to benefit from this book? From my experience, many more people will benefit than you probably would imagine. My first book, based on factory productivity, has been read by a prison unit manager and a social services manager who works with drug addicts. They both found procedures that could be applied to their jobs. I suspect the prison had a more captive audience.

I have worked in teams with company owners, vice presidents, managing directors, board members, general and other managers (e.g., production, maintenance, warehousing, quality, test, stores, planning, and design), engineers, scientists, operators, warehouse personnel, forklift drivers, and cleaners. I have deliberately gone overboard with the list. All I needed to say was that everyone benefits, but I wanted to emphasize the point. Indeed, one of the key features of Lean Manufacturing is that the processes work best when as many people from the company as possible are involved. Knowledge is drawn from across the organization, and cross-functional teams are recommended to find and fix the issues.

Adapting to New Techniques

As with any team, there is always a need for a degree of standardization; otherwise, everyone does what they want, and quality can suffer. I currently follow documented training courses, but often the team has needed to adapt the sources to suit new circumstances. Also, owing to limited time with some companies, it is impractical to try to implement complete processes such as total productive maintenance (TPM). I still train employees, though, in some of the best modules and ideas, such as 5S workplace layout, single-minute exchange of die (SMED) for quick changeovers, overall equipment efficiency (OEE) for equipment performance, and cost and consequences. The main difference now is that I need to provide staged results and cover more subjects. Thus I apply the specific techniques in the order that the *customer* most needs them, not the process, while keeping the costs as low as possible.

Most improvement books will tell you not to cherry-pick parts of the processes. They recommend implementing the processes in full to get the best results. This is good advice and should be considered, particularly if your company is planning a full implementation. I would suggest, however, that you also consider a more targeted option, too. I don't think that I cherry-pick as such. I do pick the *best bit* of a particular process, that part that I believe will solve the customer's immediate problem, but I still plan to build on what I do and develop a complete system in the future. I guess that this is like painting all the rooms in a house: It doesn't really matter the order in which you paint them as long as they are all painted eventually. Time is frequently a limitation. It can take five years to implement an improvement process such as TPM completely, and this is not an option when a company is willing to commit to only a small number of days. Ten project days limit a project to around three months. Thus I would hope to be assembling a jigsaw of processes, one piece at a time, that eventually will grow to become a more complete picture.

My experience has shown me that blending techniques is extremely powerful. I have a few favorites. This is the best way that I have found to quickly show a company what improvement techniques can do for it. I have to prove the benefits to the company team and to the management, and in addition, I need to win the hearts and minds of the employees. The more time available to explain the improvement options available and to involve more people, the more embedded the culture will become. If the team has a quick win, ideally, a production improvement or cash savings, the managers will start praising the team, and other employees will start to become interested. The company can take advantage of the growing enthusiasm and extend the project, calling in specialist trainers as required.

Diagnosis

I found that I needed to improve my diagnostic skills to enable a faster initial assessment of companies. So what works for me?

1. To help improve the profitability of their manufacturing processes, I need to find out what problems the company is facing. What are the company's main issues?
2. Next, from the list, I determine the relative impact of each problem, and then I prioritize the best issues to tackle in the time available.
3. Then I train the team(s) in the processes needed at each stage of the project. These are not intensive training courses; I just try to make them around an hour or two. Any complexities or special modules should be introduced as they are needed. (Please remember that I am talking about improvement methodologies, not safety issues, which always have to be anticipated and action taken to ensure that the operations are safe.)
4. Then I seek buy-in from the employees.
    a. I start to build the new culture. By this time, the managing director should be actively supporting the projects, but not all his managers will want to be involved, so they need to be won over, too.
    b. Then I strive for a quick win. The project has to be seen as positive by the team. It should help employees by improving their work area and/or their productivity with no extra effort. As a natural outcome of the improvements, financial savings also will occur, which will interest some of the more resistant managers.

Resistance to First Projects

When the first project is successful, the team normally wants to carry out another one immediately. Successful team members are enthusiastic and want to do more. Some will be happy to be trained to lead their own teams, which then will result in a broader implementation and more flexibility from a production perspective. As different problems are discovered, new techniques can be learned.

The hard part is *completing* the first project. I have had almost nine more years of experience since writing my first book, and I have worked with some very gifted people, both customers and colleagues, from whom I have learned a lot. Yet some things are still taking a long time to change. Considering all the different companies with which I have worked, unless the barrier really is cost, I am still surprised at how many of them have had to be persuaded to try *proven* improvement techniques. Convincing companies of the advantages of these techniques is essential.

If a company has never tried improvement initiatives, I have to prove the benefits from scratch. I also have found managers who have little trust in consultants. Interestingly, this was one of the reasons I opted to work for a government agency: Personal profit was not my goal. To be fair, just as you and I are careful about what we spend our money on, companies also have to decide how best to spend their limited cash and resources.

The practitioner/project option is based on the promise of improvement. The company has hidden costs in its provision of people, who normally would be making product. The practitioner/project option further depends on the company's own project team completing any agreed tasks and their managers freeing up the time for the teams to work. Compare this with the option of a new machine that will replace an unreliable tool and will provide positive improvements—at least for a while.

For those of us who understand equipment reliability, the improvement resulting from a new machine can be short-lived. If the old tool is unreliable only because it has not been operated properly or is not maintained, then the new machine is destined to suffer the same fate of unreliability. I digress a bit, but coming from an equipment background, I still find it a problem when a company treats equipment like fuses. Such companies run the equipment until it blows and then lose hours, days, or weeks of production time. They can't see that they have limited control of their productivity: The equipment decides when production can be run.

I mentioned the similarity between processes: Both TPM and Lean will come to many of the same conclusions about problems in a company. The "7 wastes" will show that every company needs to have a basic process that, at a minimum, inspects equipment and, better still, maintains it. In order to establish the actual condition of a machine, it is necessary to take some time to check it out. The same holds true for companies. They might be located in a "smart" building and even be making a profit, but is the company making as much as it could? Is the business effective? Like equipment, a company also will benefit from regular inspections and routine maintenance.

So why do people not jump at the opportunity to make their businesses more successful?

▲ Efficiency improvements and cost-cutting exercises are often seen by employees as threats to their jobs—and they can be. As a company becomes more efficient, it can choose to have fewer employees or to use

any time freed up to make extra product, develop new products, or make further improvements.
- ▲ Implementing an improvement project should not be done as a whim. It takes time, effort, and commitment to ensure success. It is a bit like getting a family puppy. You get immense fun from playing with the puppy and all the benefits of "man's best friend," *but* there is a downside. You need to get up early to feed it, walk it in all weather, bathe it, clean up its poop, pay vet bills, clean hairs off the furniture, and accept that the puppy sees shoes, furniture, and walls the way we see food. Fortunately, as with improvement schemes, the disadvantages reduce with time.
- ▲ Managers responsible for productivity often see anything that uses up their time for anything other than making product as time wasted. Therefore, either they don't participate or just pay lip service to the project. In the worst cases, they can actively sabotage it.
- ▲ Managers, managing directors, and owners who are responsible for profit can be wary of spending money on training or facilitation. This is especially true in times of recession.
- ▲ There is the initial concern of reduced production caused by the need to take people off-line to train them as well as to work on projects. In fact, the cost of a project is not only the price paid to the vendor. This is why the implementation has to be planned and managed properly. Companies get into a cycle of "problem, fix; problem, fix." It is hard to bite the bullet so that after the first fix, they take time out to prevent recurrence of the problem.
- ▲ Then there is the bad experience—the previous time when the implementation failed: "We tried that before, and it didn't work." One of the reasons for the "tried that before" failure is the desire to go for only the quick hits and not back up the solution with a preventative measure or process.

It is possible to jump into an area and blitz it with changes. Money will be saved, but what happens when the project has ended? If there is no built-in mechanism to ensure that the solution sustains, the improvements will fade and the benefits will disappear. The project will meet the same end as the new machine that is not serviced properly.

When I was an equipment vendor, my team used to carry out maintenance on the equipment for our customers. We could guarantee 85 percent uptime, which was pretty good. Customers would look at their uptime and decide that it was so good that we were no longer required and would cancel the contract.

We knew that if the company did not take the time to train its own people in maintenance, the machine would be running very badly in as little as nine months. We usually were called back at this point and made our money by repairing the increased breakdowns and, when the contract was restarted, the time and parts it usually took to return the machine to its formerly reliable condition.

▲ "I did a project before, and it added nothing to the bottom line." Personally, I find it hard to imagine no improvement, but assuming the figures to be true:

- ▼ It could be an accounting issue. Some companies do not differentiate direct and indirect employees. (*Direct* employees make product: *indirect* employees support and manage.)

    To explain: Consider a production line where it takes 10 employees, including the line manager, to complete a day's production. If we increased productivity of the line by 10 percent so that it now takes only nine employees to make the daily quota and not 10, this would enable the manager to return to his or her managerial job—which was not being done because the manager was always working on the line.

    The payroll department still pays for 10 employees, so no money is saved on paper. The company does not recognize that it now has a manager that it did not have previously.

- ▼ Productivity improvements often only save time. It is up to the company to ensure that the time savings are used properly. Consider a company that makes 10 units each hour. In a 10-hour day, it will produce 100 units. If we increase productivity by 10 percent, we now make 11 units each hour, so in the 10 hours we now make 110 units.
    - This is 10 more units each day—equal to one hour's production, as measured before the improvement. How do these 10 extra units make money?
    - The company must be able to sell them. This is the first and most obvious option. If the company does not sell them, it will lose even more money owing to the added cost of the materials and facilities.

- The company can save an hour's wages per employee. The company can make a day's production without any overtime.
- There is a wage saving if the shipping people do not have to stay late to ship out an urgent delivery.
- The daily orders now go out on time. There is no need to pay for air freight or a special courier.
- The company can reduce prices and possibly sell more.
- The number of employees can be reduced. If the company has 10 employees, one can be let go. This is the option that causes most resistance to productivity improvements. Sometimes, if a factory is in trouble, there is no alternative but to have layoffs. Better to save some jobs than to lose all!
- The time can be used for further improvements. The company can use the saved time to develop a new product, carry out maintenance, run a new project, cross-train employees, or train employees in new skills.

Those who do not agree with the project also can become the cause of its failure. Not all company managers want to be active participants. Great care must be taken to include them in what is being done. Avoid any suggestion that their actions are to blame for any problems found—nothing is personal.

There are also companies that believe that they can have improvements "done to them"—like having a washing machine installed. This is similar to a blitz that has not created a method of sustaining. It is also unlikely to last. These managers do not want to lose any production time and so will not release any personnel. This is the paradox. While firefighting, there is no free time, and if there is no free time, no one can be released to work on improvements. If no one is free to work on improvements, no time can be freed up. The cycle continues. The visionary manager sees beyond today and recognizes the need to take decisive action and break the cycle.

I am trying to develop a way to have a minimum-impact intervention—as yet with little success. Unfortunately, I have found that the smaller the impact to production, the longer the project takes. In addition, because fewer people are involved, the culture does not develop as well as it should. There is always the option to hire a vendor to lead an improvement program, but the vendor would need to be present on every day that project work is done, and that could be expensive. In addition, the company still will need to supply someone from the workforce. If the vendor does the job, we are back to the situation where Lean "is done to" the company.

Then what happens when the expert goes away? For the process to survive, we still need a planned successor to lead the team. We also need the team and the managers to promote the culture of improvement, which is not easy if they don't know what actually was done! And the process must be embedded, with scheduled periods for employees to participate. Otherwise, it is likely the improvements will fritter away.

I attended a seminar at Strathclyde University a few years ago. The keynote speaker was Professor Peter Hines from Cardiff University. He explained his analysis of the benefits of short, high-intensity blitz interventions compared with the long-term, measured introduction of an improvement program. I hope my understanding is accurate. In short, he demonstrated that for an improvement program to be sustained,

- ▲ All the employees must be trained properly.
- ▲ The improvements must be embedded in daily routines.
- ▲ The culture of the company must include the concept of ongoing improvements.

If a blitz is carried out properly—and there are consultants who do it very well—it is essential that follow-up steps are included. Making the changes is only half the job. Ensuring that the changes survive is the other half. I think this is the third time that I have made this point: For lasting success, it is essential that the improvement work is *not* regarded as an extra job to be carried out when normal work is complete. The improvements must be ongoing. A framework or timetable should be developed to ensure just that. Please remember, however, that if the company ignores the sustaining measures, the project will whimper to a stop.

Where Will We Get the Time to Do Projects?

This is a true story. A production manager once told me something his predecessor said: "If we make all the product right the first time, what will the rework lines do?"

Several points immediately spring to mind, but let us just consider the big ones:

▲ This ex-manager had no idea how much it cost to make his product.
▲ He had poor (or no) documented processes.
▲ He had poor (or no) quality control.
▲ He had not tried to improve the process to make it more efficient. There was no need: The rework line would pick up the problems.

This manager's successor told me another story, too. The product the company made was physically big. It was bigger than a stand-up refrigerator/freezer. It had a large number of components, many of which were bought as subassemblies from the company's suppliers. One unit had 10 screws that held it in place on the main assembly frame, and the production line fitted one of these subassemblies early in the process. However, it turns out that the assembly teams often fitted only three of the 10 screws. Why would they do this? Well, it seems that the incoming units rarely worked, and fitting only three screws made it easier to remove the faulty unit later when it failed at the testing stage.

I feel patronizing even trying to explain what the company should have done here. There should have been an incoming quality standard for the subassemblies. There also should have been appropriate testing before installation. The number of failures should be tracked and charted to assess supplier reliability. Ultimately, if the supplier continues to be unreliable, it needs to be replaced. As we do in our mapping evaluation, it helps to estimate how much a failure costs to put right. This helps to offset any argument that the service is acceptable because the supplier might be the cheapest.

If we plan to run a continuous improvement program, we must find time for the team to do the work *and* minimize the impact on production. The preceding is just one example of extra work being done that wastes the time of employees. When any company does the same job more than once or has the same failure more than once, it is basically paying twice. All the extra wages and costs are a straight loss and add to any reduction in profits. We need to avoid the time wasted and use it more productively.

What time is there to be recovered? We have wasted the time of the production staff, the testing department, the repair department, and the shipping/warehousing department. There also will be an impact for administrative paperwork. It also wastes the time of the original manufacturer who has to fix the returned units. (That manufacturer clearly needs its own improvement plan.) Over and above the basic labor costs, we have costs for electricity, heating, delivery, telephone calls to the supplier and the customer, paperwork, and packaging, and because the product has to have extra work done, the unit takes longer to complete. This could lead to overtime costs and late deliveries, with extra air freight or overnight delivery charges.

So, to be a bit more accurate, our profit equation should be changed to

$$\text{Profit} = \text{turnover} - (\text{the cost to make the product and sell it} + \text{the wastes})$$

The cost to make the product and sell it is technically correct by itself. The real problem is the *waste* component. It is not an essential spend, and it is not normally recognized as being unnecessary. The wastes are automatically accepted as being a normal manufacturing cost. If the waste is not recognized, then there is no reason to eliminate it. The ideal production line would make the product *right the first time*, and there would be no need to have a rework line.

Did I mention that the company no longer exists?

Productivity Tip

The number of products that fail at each stage of a process should be monitored and targeted for improvement. This is not intended to shame the operators but to identify whether a process improvement is needed and quickly show when something has gone wrong with the equipment or the materials. The improvement process is known as *short interval control* (SIC).

To embrace employee involvement, the data should be displayed on a chart or graph to be updated by the assembler or operator—not the team leader. The chart should be placed on a notice board where the results clearly can be seen by everyone. If the failures are clearly visible, it is openly recognized as an area that needs improvement. However, if the graph is located in a directory on a PC, even if people know where it is, no one but the production manager is likely to see it, and the failures will become just another statistic for PowerPoint presentations.

A second example of a time savings was seen in a company where thousands of incoming goods were sorted into different types and then counted. This was an essential step because it was what the customer wanted. The team and I used a map to review the process and quickly discovered that the next step in the process was for each unit to be bar-coded and scanned. The map showed that counting took three days, and the scanning took a further four days. Counting and scanning are virtually the same job, and in Lean, doing the same job twice is a recognized waste.

The three days to count was a customer milestone that had to be met. The team decided that if we added an extra workstation to the scanning team, we could reduce the time to sort *and* scan to three days. With a minor change, the software could count the units. This would cut the overall process lead time by a whole four days, which then could be used to make further improvements and/or extra product.

I will point out here that you don't need to wait for the project to be complete before implementing positive improvements. If you are certain no issues will be created by the changes, you can make the improvements and get an immediate return. The time savings alone will benefit the team.

A third example involved a company that used a team of engineers to create customer quotations that included complex technical drawings. It could take several days for the drawings to be created. Once the quotation was complete, it was sent to the customer. If the customer accepted the quotation, the job was passed to the production teams, which started from scratch and re-created new drawings—wasting several days.

A fourth example helped a company that ran a process that used 10 automated machines. The equipment was routinely shut down at lunch and breaks, even though it was designed to stop in the event of a failure. This resulted in the loss of at least one hour of production each shift.

As it turned out, the equipment was very reliable, and the process did not require constant supervision. By simply leaving the machines running and responding to any alarms, the company increased productivity by around 10 percent and made upwards of $100,000 of extra product. In this instance, the extra time was used for production, but it could have been used for training and improvement projects.

Overall Equipment Efficiency

A fifth example represents a different type of improvement. It is from TPM and a module called *overall equipment efficiency* (OEE). It can be used as a stand-alone process.

OEE = performance × availability × quality

Basically, any piece of equipment has a capacity. It should be able to make a specific number of units in a given time. Let's say that it can make 10 units in an hour. In a 40-hour week, it should have the capacity to make 400 units.

Its *performance*, when everything is perfect, should be 400 units per week. However, if the machine is not running properly and has to be slowed to half speed, it would make only 200 units and not 400 units.

Performance now would be (400 − 200)/400 = 50 percent = 0.50. Performance also is affected by operator skill, quality of the operator's instructions, and availability of materials (stock).

In the fourth example, we suggested that we had equipment that was shut down for one hour each day for breaks. Thus, over five days, the equipment was not "available" for use for a five-hour period. In a 40-hour week, the percentage *availability* is (40 − 5)/40 = 87.5 percent = 0.875.

With the lost time, we can only make 0.875 times 200 = 175 units in our week.

Availability tracks any time the machine is not available for use. This can be due to no operator, downtime, maintenance, changeovers, and setup.

Its *quality* is the percentage of total units minus faulty units divided by the total number of units made. Thus, if we have four failures every week and 175 units have been made, the *quality* would be (175 − 4)/175 = 97.7 percent = 0.977. Quality is also affected by reworks and the standard of the parts used.

OEE = performance × availability × quality
    = 0.5 × 0.875 × 0.977
    = 0.427
    = 42.7 percent

An OEE of 42.7 percent, as shown above, gives us only 171 units in a week, nowhere near the perfect 400. This is a clear guide that something needs to be fixed! In reality, an OEE of 85 percent is regarded as world class.

One more complication is the never-ending debate on whether OEE should be based on 24 hours over seven days or just the number of hours the machine is operated. The percentage OEE is higher if we only consider the shifts.

For simplicity, if we have a perfect OEE of 100 percent, but the factory is only open for a single eight-hour shift five days each week and the company shuts down the rest of the week, do we have an OEE of 100 percent?

What about the 128 hours the machine is not available because the factory is closed? Should the availability not be only $40/168 = 23.8$ percent?

I use both—the first because it makes the team feel that the performance is better, and the second because it shows much more clearly that almost 76 percent of improvement is possible. In our case, if a huge order appeared, we would know that there are 128 extra hours of availability, so we could take it with no problems—other than personnel.

I will rarely assume that the reader has previous knowledge. I will discuss the key areas of production, including efficiencies, product quality, reworks, throughput, downtime, planning, targets, the universe and everything. *But* your skills will increase quickly as you progress through the book. The savings will come soon after.

Culture

Company cultures vary immensely. I guess my main observation is that the culture reflects the style of management. The downside is that the reverberations of bad managers can last for years. Even if a "bad" company improves, people have long memories. One employee told me about the company's "hire and fire" policy. When was the last person laid off? "Ah, Bert was fired eight years ago. . . ."

I try and get a feel for the company when I make a visit to assess its issues. There are some obvious signs of a good company. The managers call people by name as they walk through the plant. People feel that they can talk to the managers. People do not constantly complain about the place. There is a good, friendly atmosphere. The operators know what they need to do and have good operating instructions. They feel involved in the way the company runs. Managers talk to one another. Employees feel secure and do not complain about their wages all the time. There are other clues, but these are the ones that spring to mind. There are some variations based on the type of company, but even the variations are quite diverse.

Family-run companies are often very successful. Frequently, they have employees who have picked up their skills from relatives and parents. Family members tend to work for the company. In fact, that is often the reason for the family business. It is often a lifestyle choice.

Although I am stereotyping a bit, most family-run companies have a friendly culture, but some managers keep a tight reign: They can treat their children at work as they would at home and not as company managers. This can spill over to the way they treat other employees. I have found a tighter control of spending because they are spending their own money. I guess I was like that, too, when I had my own business. In the smaller companies, key jobs are held by family members, who often have had only on-the-job training. However, unless increasing turnover is *not* a major driver, as is often the case with "lifestyle" companies, as the company grows, there is an increasing need for family members to become more professional.

The skills that have proven successful in the past are assumed to be good forever. This can be applied to all companies, but owing to a slower management turnover, it seems more prevalent in family-run companies. Many of the established family-run companies now ensure that family members have been educated to degree level or to an appropriate standard in their specific areas.

Analyzing the company processes (see below) will identify what the current problems are and, among other things, where any training is needed for all employees. Mapping is a perfect technique to highlight training issues in such a way that no one is offended.

Medium-sized companies have another culture—medium being defined by the U.K. manufacturing support organization Business, Enterprise and Regulatory Reform (BERR) (formerly the Department of Trade and Industry [DTI]) as companies that have up to 249 employees. As in family-run businesses, growing in size exposes issues. Some of them have a few bridges to cross to take them from being informally run companies to having proper, formal procedures. Product reliability becomes more important as the products become brand names with an expected level of quality, and employees need proper development. Thus such companies set up human resources (HR) departments, quality checks, skills reviews, strategy planning, and training for managers.

Large companies, often global, have a third type of culture. They have established brands. Their customers have expectations of their products. Specific standards of service are expected. However, being large does not guarantee that a company is efficient. What some smaller companies would see as a major loss often can be seen as acceptable in large companies while profits are flowing. It is often the case that large companies are not even aware of the losses, but I think that this is also true of all the other types of company, too.

I feel a need to clarify my views on formal processes. I would never advocate military-style management with barely achievable work quotas and little job satisfaction. I do, however, promote targets as a means to track progress and identify problems quickly. I believe that it is wrong for targets to be used as a club to keep the workforce on its toes. In fact, bad targets breed bad working methods. To quote Agnes Pollock, "Targets breed behaviors." Let us consider a target of 10 units a day. If time is a constraint, corners might be cut during manufacture. This could mean that we will achieve the 10 units we want, but only 5 of them might be sellable. We have achieved the "target," but we have cut the quality, so in the long term, we have lost. If sales numbers are a target, we can get "mis-selling," as alleged in relation to some recent door-to-door energy suppliers. This leads to a series of complaints and time wasted putting things right—to say nothing of the company's lost reputation.

My goal is a process where the same task is always carried out the same way using the same materials with the steps in the same order and assembly to the same standard. If this is the process, the product always will be correct the first time and built to the same quality standard.

Why Is It So Important That the Initiative Works and Is Sustained?

When companies are thinking about taking the first leap of faith, you might hear doubts such as, "We tried this before and our project did save us money, but then it all disappeared."

There are procedures that will help you to find and pick the low-hanging fruit. These help to find and solve the easy issues. They will save you money—fast. In my experience, this is not always a bad thing. Even if you just find the obvious losses owing to the "7 wastes," this will generate the cash and provide the time to develop a lasting program.

The quick hit has its place, provided that the people involved understand that a quick hit is all they will be getting if the process is not continued. It can be used to convince unbelievers, to motivate the teams, and to prove that there are real benefits in doing the work. However, just to make the point once more, if the procedures are not embedded in the culture and the process, the lack of sustainability will reinforce every "I told you so" argument that was used before the implementation. This will make getting buy-in from your employees for any future projects even harder. And last but not least, you will lose the year-on-year benefits you would have made had it kept running.

How do we find out what needs to be fixed? Chapter 2 will discuss a few scarily simple techniques to identify the opportunities that will benefit your factory or process. Culture change will be discussed as it arises.

First, a couple of apologies. A lot of my experience is in problem solving and training, albeit in universities, hospitals, electronics manufacturers, and a host of other industries from food to textiles to engineering. With a science background, I tend to be logical and sometimes provide too much detail. Experience has taught me that it is better if you ask, "Why is he telling me that?" than that I omit something that might help. I am also not as politically correct as I should be, but I will try to do better. Oh yes, and I have an odd sense of humor. I guess that is why I like training different groups of people: In this environmental era, I get to tell the same jokes again and again and call it recycling.

CHAPTER 2

Finding Improvement Opportunities

It is safe to assume that every company has something that needs to be fixed. The managers of each company will be aware of some of the issues, but there will be others that are not even recognized as being problems. What is less likely is that the managers will know how much any of the issues are costing them. If they did, they would definitely do something about them.

I am a convert. I now actively promote mapping as the best method I have ever used to diagnose issues in a company. In this chapter I introduce a few maps that I use for problem finding and process improvement. In Lean Manufacturing, I've found that the most commonly used maps are process and big picture maps (each explained in detail in its own chapter). It is strange that the most used map is not the value-stream map (VSM), but I would not be surprised if others use it more often than I realize.

I also will consider how maps evolve. I developed one to solve a particular issue, but because I found it so helpful, I developed a couple of other add-ons, too. I have called one the *capacity map* and the other the *value- and capacity-stream map*, which is based on the value-stream map. Neither of them is a new technique. I blended mapping with the theory of constraints and overall equipment efficiency (OEE) so that I could highlight some added value by pointing out an inefficient step.

As always, I will not limit the content to Lean Manufacturing. I also intend to include my favorite techniques from other methodologies. After all, Lean, like all other improvement processes, is constantly evolving to suit the needs of its users. I have included one nonmapping process, based on simple brainstorming. I use it in my problem-solving course. This is not a new technique either, but when coupled with Lean and losses, it is a very good way to quickly quantify issues. It is not as detailed or beneficial as mapping, but the advantage is that it takes only a few hours to complete and is an excellent platform to introduce the *need* for a proper, factory-wide improvement program. At the end of the process, everyone will have a (depressingly?) more accurate idea of what the issues are. But more to the point, the team will appreciate how much the issues are costing the company. Armed with this knowledge, it will be difficult to deny the need for formal improvements. This approach also reinforces the positive argument that the new process is not about cutting head count.

Adding *Some* Value?

There are production managers who believe that there is no problem as long as product is coming off the end of the line. Making fewer products than yesterday, last week, or the week before is never a consideration. These managers are happy as long as the pallets or boxes eventually fill up. How long it actually takes to complete an order is not relevant. After all, there is always overtime to make up the shortfall, and they have never heard any customer complaints when the order was only a few days late.

This type of inefficiency is why I started to consider overall equipment efficiency (OEE) as well as the Lean options of value-added or non-value-added. OEE is a measure of how much a process *does* make as a fraction of how much it *could* make. For example, I will tell you the story of a machine consisting of two key sections. The product is processed on one tray while a second tray is loaded with raw materials. After processing, the two trays are exchanged. The processed tray is now unloaded and refilled, while the tray with raw materials is being processed. Easy!

The tray exchange should take 30 seconds. Thus, if the processing takes 90 seconds, we should be able to process 30 trays in an hour (60 minutes divided by 2). However, the mechanism that loads and exchanges the trays requires a precision setup to work reliably. One company I visited found another way: It slowed the speed of the tray exchange down to "six million dollar man" speeds. It was like watching the operation in slow motion. The exchange now took almost three minutes—but it worked perfectly every single time (Figure 2.1).

The product throughput now was limited by the loading equipment. Each process run now had to wait a full minute for the tray to be loaded, and the company could process only 20 trays in an hour, not the 30 of the other

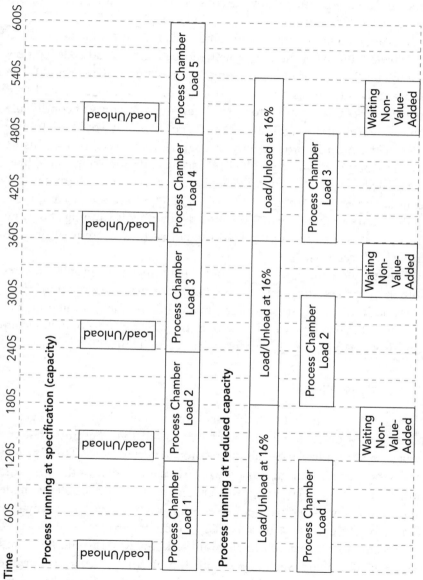

Figure 2.1 The tray load/unload and process sequence.

company. The productivity had been reduced by 33 percent. The equipment engineer was delighted because the tray change was more reliable and needed no attention. But did anyone consider the production loss? No. Did anyone realize how much money was being lost?

In Lean, loading the tray is still value-added. We must load the tray to enable processing. After the adjustment, the load and unload cycle has now become a bottleneck because it is taking six times longer than it should. The loading now introduces an obvious one-minute non-value-added step where the process is "waiting" to start. Unless the company was continually monitoring overall throughput—which it should be—it would not be able to detect the drop in productivity. Recognizing this type of issue using normal mapping is quite difficult. The loading and processing would have to be treated as a parallel process as opposed to a serial one. Alternatively, if the modules were measured in terms of capability based on the equipment specification, we would see that the process chamber is running at only 66 percent of its capacity and the load/unload cycle is running at only 16 percent of its capacity.

This is a good example of a situation where curing one problem creates another. It makes the perfect case for monitoring production rates. However, if the adjustments were made before the production was monitored, the baseline would be wrong. Tracking this type of inefficiency is one of the reasons for my developing the capacity map.

Right First Time

As a customer engineer, I quickly discovered that all I had to do to get an immediate 10 percent improvement was to tell the operators that the managers were really worried about the drop in performance. Then I would stand beside the machine and assess what was happening. For the operators and engineers, the very thought of anyone looking for problems made things run better. Watching them, like using graphs to track performance, seems to be a trigger for higher standards of operation.

It occurs to me that you might want to know why I would just stand beside a machine and watch. I was called in by companies to fix specific issues or breakdowns. I would be given a few symptoms over the phone, but more often than not, the company had little idea what was wrong. If the company was in another country, I had to use the initial information to make a best guess at the problem and arrange any technical information or parts that I might need to take with me.

On arrival, my first objective was to diagnose the *root cause* of the problem. If the machine had been down for a while, the solution was rarely just the original issue. It is normal for local engineers to carry out first-line repairs, so it also was important to find out what they had already done. Their repairs unwittingly could have introduced new problems that also would need to be resolved. A changed circuit board, for example, might need to be calibrated.

It was important that I made a correct diagnosis and repair. The longer the total downtime, the higher were the production losses. If I left at the end of the job and the problem was not fixed properly, apart from having a furious customer shouting at my boss, it could take a long time to get back—particularly if a couple of flights and a car journey were needed to get there. *Right first time*, a Lean and total productive maintenance (TPM) goal, also should be a key objective for every vendor engineer.

To help make the most accurate diagnosis and determine all the symptoms, I spoke to as many (involved) people as I could. Then I would watch the machine (or process) for myself to see it running. The Japanese talk about the *Gemba*: This is going to the source of the problem to see the issues firsthand and not to rely on a secondhand report. If you don't know how a process or machine should run, it is very difficult to tell what isn't working. The fault symptom we see is not always the real cause of the issue; often, the root cause is several steps earlier in the process.

Example: Root Cause

As an illustration of the cause not being where the symptom is seen, I will tell you about a joinery company with a process step involving sticking materials to wood. For days, even weeks, there would be no problems, and then, suddenly, there would be a series of failures. Often the problem would go away by itself, but it always came back. The company was highly professional. It checked everything: The wood surface was flat; the presses worked, and the faces of the presses were undamaged; the heaters on the press were working; the press temperature was correct; the material and the wood surfaces were clean; and the glue was spreading properly.

Naturally, a new tin of glue was tried every time. The company made sure that it had extra tins in store. The company even made sure that the operators were trained properly. Yet, despite all those checks, the problem persisted. Everything pointed to the glue. When contacted for advice, the glue supplier sent a representative with a known "good" batch to prove that the glue worked. It did. The glue was confirmed as the problem: Issue solved. It must have been a bad batch—another?

Then the problem returned. Only this time the company knew that the glue was good, so it delved even deeper, but with no success. In the end, the real issue did turn out to be the glue, but it was due to the way it was stored—not the glue itself. The glue had a minimum storage temperature as well as an operating temperature. As it turned out, the company's store had a concrete floor. Overnight, when the heating was off, the temperature *sometimes* fell below the minimum temperature for the glue. This explains why the issue was intermittent, lasted for varying periods of time, and why not every tin of glue failed.

I am not sure whether a general process map on its own would have shown this to have been the cause. Storage conditions for individual products may not have been detailed at this point. However, such a map definitely would help in a problem-solving situation. When we are looking for a cause, every step of the map would be under critical scrutiny by the team, and every process step and specification would be included.

Size Isn't Everything

The fundamental point I want to make is that simply by taking the time to look for issues, you definitely will find some. Every time the process is repeated, you will uncover new problems. Conversely, if you don't look, improvements will need an external incentive to get them started. Nothing will improve until there is a serious problem—one that forces immediate action. Even then, if the problem is not understood fully, the solution may not be permanent.

Serious issues usually cost a company a lot of money and have an immediate impact. However, not all the expensive issues are big ones. Small, recurring problems virtually always tend to be overlooked. This in itself is a fundamental mistake made by many manufacturers! In TPM, issues that frequently repeat are known as *minor stops*. In one company, when the annual impact of a recurring problem was analyzed, it was estimated to cost $100,000 a year—each and every year.

I tend to find that to force a full-blown repair, the failure often needs to be

- ▲ On the scale of a production batch that has been wiped out
- ▲ Failure to pass a customer audit
- ▲ A customer who takes or threatens to take her business elsewhere because another company offers
  - ▼ A higher standard of product
  - ▼ A lower price
  - ▼ Better delivery dates
- ▲ Or, in the case of poor operation, an equipment breakdown

Customer Audits

Failing a customer audit is becoming more prevalent. One reason for taking business away is the "perception" that a factory has low standards or concerns about negative publicity. I have experience with a couple of companies where business was withdrawn for this reason. Many companies now routinely audit their suppliers. This includes visiting companies they have dealt with for years. If the appearance of the factory looks bad, it suggests that the company doesn't care about quality. In both companies, 5S improved the appearance and layout of the factories.

Where Do We Start?

Only 60 percent of production issues are caused on the factory floor. The other 40 percent are based in administration. Ordering a part too late or not at all also will stop production. So, too, will not being able to find the material needed. Therefore, returning to the first question, how do we actually start the project moving?

Whether you know what the issues are or not, the mapping process should be followed. To get a general appreciation of the problems, have an informal talk with all the departments. The whole company is involved in making product, not just one department. Involve as many people as you can, and logically work your way through the operation.

You might begin by focusing on a single area, but you shouldn't stop there. The product has to *flow* smoothly through the whole factory. Flow is one of the five Lean principles. Each department (or process or machine) gets fed from the previous one. It then feeds the next workstation. Production at each stage depends on getting exactly what that stage needs from the previous stage. What each stage needs is the right number of quality units of usable parts when it needs them.

The Five Principles of Lean

Lean is defined from the perspective of the customer, not the manufacturer, but both parties gain from its application. As an introduction, I have listed the five principles of Lean below:

1. *Value.* Value is something that *adds* to a process or to a customer's satisfaction levels. Added value is good; non-value-added is bad. Something only has value if the customer is willing to pay for it.

    In addition to the material cost and any equipment needed, the price of your product is based on how long it takes to make. But what if your production line takes twice as long as your competition? Possibly you do not have an organized process, perhaps you do not run equipment to its capability, or maybe you have poor quality control. The customer will not want to pay for your inefficiency.

    It is important to understand what the customer wants from your product. If you can supply what the customer wants as opposed to what you think he wants, you will have a happy customer. For example, adding multiple layers of packing might not be what the customer wants. Perhaps he has to pay someone to unwrap the product and put the wrapping in a bin. He also may have to pay to dispose of the wrapping. To the customer, the packaging has no added value. Indeed, it costs him time and money, so you both lose because you are paying for it too.

    There are some essential steps that do not add value as such. These tend to be quality checks, mandatory inspections, and safety procedures. Poor process standards can lead to a need for unnecessary quality checks, though.

2. *Value stream.* This represents every step from the start to the end of the company process. It essentially looks for waste. Every step of a process should bring the product closer to completion. Each step should *add*

value to it. Value is not always monetary but should be something the customer wants.

Conversely, every time the process stops and has to wait for something, we are not adding value, and the stream has been interrupted. Of the "7 wastes," waiting creates some of the biggest losses.

Let us consider the packaging example. If the customer doesn't want the packaging and the product doesn't need it for protection, then these steps in your manufacturing process do not add value for either of you. You are paying for unnecessary materials and for someone to do the work. In addition, if operators are working on the packaging, they are not working on the steps that do add value.

3. *Make the product flow.* Product should move smoothly through each process step. Every time the product comes to a halt or needs extra work, the flow is interrupted and starts costing you time and money.

Good flow is what you would like a perfect car journey to be. No bad weather, no red traffic lights, no road work, no queues, no parades or trucks blocking streets. In short, you want no interruptions. Any time you spend waiting or on detours does not add value to your journey. If you are traveling in a taxi, for example, the meter keeps running when you are stopped, and you pay for all the delays and the driver pays for the extra fuel.

4. *Pull.* Make the product at the pull of the customer. *Pull* is customer demand. The perfect goal is to make only what the customer wants. If you have to guess how much product is needed, there is a chance that you will make more product than you can sell. The extra materials and labor used by making more than the customer wants does not bring in any money—they have no added value. There is additional waste in storage costs, material control, and for scrap if no one buys the extra product.

Trying to control what you make relative to sales is the objective.

5. *Strive for perfection.* This is a goal in every improvement process.

But, for me, practicality and limited funding suggest that perfection should be a longer-term goal. My belief is contrary to key principles of a range of improvement techniques. I believe that the benefits of improvements must be weighed against the returns, and the issues that provide the greatest benefit should be prioritized. For example, TPM recommends that costs are not considered because doing so will prevent the smaller issues from ever being resolved.

How Do We Do It?

If I had never used a technique before, I would start with a book to get a basic understanding of what to expect. Engaging a consultant is another option and is a good one. Cost can be an issue, so make sure that you can afford it. I and many of my colleagues have found support costs to be a barrier in some companies. It could be argued that I am a bad salesman, but I would disagree. Could I be too honest? It can be a hard decision for a company to spend money on the "chance" that a savings will be made. A company will be hard pressed *not* to get back at least what it spent. Indeed, in my own experience, any savings will be much, much more. The outlay has to be recognized as the investment it is—no different than buying a new piece of production equipment or a software package or even starting a new employee. If you choose to go with a consultant, interview more than one company. Ask for references. Compare what is offered. You would do no less if you were buying a television.

A consultant possibly will review your company (free of charge) and then recommend what she thinks you need. You probably will find that what you need will vary from consultant to consultant, just as would any estimate for a job. There is a natural tendency for people to see problems in the areas in which they have the most experience. I frequently hear the statement, "If you have a hammer, you will always find nails." Personally, I don't think vendors are that shallow. They will see most of the "nails" in their area of expertise, but they also will see the other opportunities. After all, the more "nails" they find, the more work for their hammer!

Make sure that the consultant has experience in all the areas she offers or that she has a colleague who is able to provide the support you need. If the consultant makes her living as a consultant, is she busy? Consultants make their money from the number of days they sell.

Make sure that you are getting training at a pace that suits you. I had experience with one company in which the vendor offered a training rate *much* slower than I would have given it. The company was delighted with this rate because it was not too demanding on its employees. Fortunately, the company was a major organization that could afford the extra training cost. Conversely, I have worked with companies that have had to reduce the frequency of training days because they were a drain on their productivity and they did not have the personnel to cope.

When reviewing the company, time permitting, a good consultant will talk to a range of people across the company from the managing director to the warehouse personnel. In this way, she will get a feel for the issues that prevent all the employees from doing their daily job. It also will show whether management is in tune with how the factory works at the production level. For any program you are offered, find out how long it is likely to take before you see returns.

The duration of the project could need revision after it has begun. It is possible that unforeseen issues will be uncovered during the work. It is well worth having a project review partway into the project to confirm whether the original plan is still the best option. Changes may be essential if the new issues have a bigger impact than previously thought. Nevertheless, you must get a feel for what the project will cost you in time, production, and labor. Compare your input with the return expected. When estimating, I believe that it is better to underestimate the benefits and be pleasantly surprised than to overestimate them and be disappointed.

Is there a government agency that can advise or help? In the United Kingdom, for example, there are manufacturing advisory services (MAS). In Scotland, they are known as the Scottish Manufacturing Advisory Service (SMAS). There are also university departments such as Strathclyde University's Design Manufacture and Engineering Management (DMEM) and Quality Scotland plus a plethora of other groups. You can find contact addresses and phone numbers on the Web. For example, www.nist.gov/mep/manufacturers/process-improvements.cfm is the address given to me for advice in the United States.

The third option is to do the work yourself. (I am not just saying this because I write how-to books!) I am not advocating tackling anything overly complex. I have learned huge amounts from books—one of the benefits of a job that involves world travel and lots of time waiting in airports. You might read a book on Lean Manufacturing, 5S, and TPM and/or get some training for yourself, even if it is just to see how the trainer teaches the class. Some government agencies and training companies run introductory events to promote their products.

It was not until I began to assess a much larger number of companies that I started using mapping. I frequently used PowerPoint-style flowcharts, which help a bit, but they are a much less sensitive tool. They tend to look at larger chunks and, if drawn on a PC, are limited by the size of the screen. This is similar to working on a very large spreadsheet: You cannot see the whole page. I learned the mapping processes from some of my SMAS colleagues. It was a revelation. I would say that it was on a par with discovering that you can buy special underwear for cycling that has gel pads fitted. Anyway, I now truly believe that mapping is the best process with which to start. I can't believe that it took me so long to find this out! The downside to starting with mapping is that the financial savings are delayed, but they will be better in the end.

However, if you just want to get a feel for the problems and how much they are costing, then, as a precursor to process mapping, try the problem-solving technique:

1. *Gather a group of people into a room.* Ideally, they should be a cross-section of people from different areas in the factory and be employees you know will participate.
2. *Split them into a couple of teams.* Ensure that people from the same department are on different teams.
3. *Give each team a flip chart.*
4. *Ask each team to list 5 to 10 problems each.* These should be issues that prevent them from doing their jobs on a daily basis.
5. *Use the brainstorming technique to get ideas.*

Productivity Tip: Brainstorming

Brainstorming is a much underused technique. Despite popular belief, it is not just an exercise that people use when they are in training courses. It should be used any time you are trying to get more understanding about an issue. At meetings, it is common to have roundtable discussions, where people add their thoughts. But the results are better if formal brainstorming is used.

The "Nail Game" is a good example of the benefits of brainstorming versus meetings. The object of the game is to drive an ordinary six-inch round-headed nail into a block of wood and then see how many nails the team can balance on its head. Most teams find that they can balance one nail and don't believe you when you tell them that balancing over a hundred is not unusual. Naturally, there is a really clever solution. Easy, once you already know the answer—but I am not going to tell you here.

The first stage of the game is to present the problem to the team and let team members discuss possible solutions. Virtually every person throws ideas into the ring, and some of these ideas are good clues to the solution—but no one notices. Nothing is recorded, so team members can't even look back and see what they said before! When they repeat the exercise, this time using brainstorming techniques, they still don't get the solution.

But the key points are now recorded on the flip chart. All I need to do is to get team members to start linking some of the ideas *by working together*, not as individuals.

Everyone knows how to brainstorm. I usually use two examples:

▲ Have you ever sat in a bar with a few friends and discussed what a football team manager or referee did wrong during the game and what you would have done instead?
▲ Have you ever seen people at a wedding discussing what other folk are wearing—and how they would look better if . . . .

Both these groups are basically brainstorming, except that the points made (the problems) should be recorded for subsequent analysis. I use Post-it Notes to record the data, one for each point raised, because they can be easily moved around on the flip chart.

The number of issues recorded is not important—the more the merrier. Indeed, the process has a natural stopping point. It is like making microwave popcorn: the "pops" will be thick and fast at the start of the session, but the process should end when you slow to only one "pop" every minute or so. The Post-its will be reviewed at the end, with the team checking to see if any of the ideas can be developed or combined to make one, better idea.

When looking for company problems, the Post-its can be regrouped into common issues and root causes sought. Additionally, each issue should be evaluated to see how much it costs the company each time it happens. Better still, estimate how often the problem repeats over a year. This will provide a standardized list that will help in decision making. The issues that cost the company the most in time and/or money and/or customer issues can be prioritized for resolution.

The rules for brainstorming are simple but must be followed if the team is to get maximum benefit from the exercise (see Chapter 7).

I normally end a problem-solving course with the teams listing issues in order of (annualized) cost to the company and include estimates of what it will take to fix them. The teams also will see if their group might be the cause of problems for others and get an appreciation of why they need to be fixed. The list also provides an opportunity to discuss the various improvement techniques available to fix the issues and, as an added benefit, begin to change the company culture to one of continuous improvement. Throughout this book I will regularly introduce the cost of problems and include a simple way to quantify the results later.

Mapping: General Guidance

Mapping, the main topic of the book, is now one of my favorite tools. It is very powerful and is surprisingly easy to do. Even if the process is not performed correctly, you can still get good results. Some of the problems you will discover may have been around for a while—often for years. It will be common to ask, "Why have they not been resolved before now?" This will be emphasized when you estimate how much the issues are really costing the company.

To make the first analysis, I recommend reviewing the whole operation from sales and ordering of materials to shipping of finished goods. Find the biggest, *solvable* issues, and prioritize the improvements. To do this, you will use *big picture mapping*. It should be appreciated that not all manufacturing processes are complicated, but they still can have big problems. One process I worked with was very simple:

1. Deliver the product to the factory.
2. Unload the materials into the loading area.
3. Transfer the materials into a processing drum with a conveyor belt at the output, checking the quality as the material is transferred. The moving belt feeds a machine that squashes the material into cubes (bales).
4. Sort the bales.
5. Store the bales.
6. Load the bales into vehicles.
7. Ship the bales to the next process.

When mapping a process, please remember that it does not to be a race. Everything does not have to be done at once. Use your personnel such that availability does not overly affect daily operations. If you try to do too much, you might end up deciding to abandon the process, which is the worst thing to do. Equally, if you do too little, the team will think that the task is not important. Involve the team in decision making. The map can take as little as a day to complete, but three days tends to be a fair average for a good-sized process.

There most likely will be some initial impact on productivity unless overtime is used to cover. Time is always at a premium. There is a need to break a cycle of inefficiency and win back some time before things start to get better. The most common example of this can be seen in companies where no planned maintenance is carried out on the production equipment. "We don't have time to carry out maintenance. We are always fixing breakdowns." (*Firefighting* is the technical term used. It makes the lack of maintenance sound professional.)

To lead the process-mapping exercise, read the relevant sections of this book. Try to gain a bit of extra knowledge about the process. This will help to facilitate the analysis and increase the team's confidence. Team members will ask the same questions as you did. Do not do all the work alone; use the team members. The goal is for the team to create the map with input from everyone.

To get the best out of the procedure,

1. *Know what your process actually is.* Reliability-centered maintenance (RCM) has long recognized that there are often two processes: the written specification and the one everyone tends to follow. Analyze the "real" process—the one that is actually used—not the specification.

    Note: Find out why the written process is *not* followed. It is likely that it contains problems or is hard to implement, and the operators have found it necessary to introduce ways to get around the problems.

    If a process works and the employees understand *why* it has been written this way, normally, they will use it. It is only when a process does not work or causes problems that people tend to avoid using it.

2. *Walk through the process to see it in operation.* This will help you to decide who should be involved in the improvement team. You will repeat this step with the process analysis team later.

3. *Talk to the operators and get a general appreciation of their issues.* These include instructions, quality, rework, materials, equipment reliability, and what stops the operators from doing their job on a daily basis.

4. *Understand why there is a need to fix the problems.* Once it is known how an issue affects the operation and how much it is costing, it is easier to get an appreciation of how much time or money could be spent on

resolving it. But please remember that spending money is not always the answer—brainpower is often better.

Are the issues wasting time and costing money? Find out how much time and how much money. Is it a quality issue? The "7 wastes" will help to identify most of the issues.

5. *Decide if you want to analyze the whole process or just part of it.* A pilot project can be used to achieve results faster. If a part of the process is chosen, use the process map, but expect it to overlap into issues from other areas, particularly administration. The improvements the team discovers may help more than one product line. The improvements can be rolled out to the other processes.

6. *Select some good people to work with on the map.* I have specified *good* employees here. These will be the ones you know will commit to the process and make it work. Their success will help to motivate the teams that follow.

   Previously, I would include a difficult employee. When I convinced him or her that the process works, this was a powerful proof for everyone else. Unless there is no option, I don't do that now: I can't afford the extra time it takes. I would include a difficult employee in a later team that includes some "converted" members for support.

7. *Find out any basic knowledge needed to tackle the job.* Is any equipment, training, measuring devices, or data collection needed? Don't forget safety issues.

8. *Get a basic understanding of what can go wrong with each step of the process.* There is no way that everything can be known. If this were the case, there would be no need to map the process at all. Read Chapter 7 on simple decision making. When solutions are proposed, remember to ask whether there are any obvious consequences with proposed actions Don't, for example, reorganize your whole factory at once. Test a new idea on a pilot line.

9. *Carry out the mapping as a team exercise.* I normally get the employees from the area to do the physical mapping, and I stand back to let the others on the team take the lead, providing guidance as required. Don't get overwhelmed by the size of the task. Remember how to eat an elephant: one bite at a time.

Before putting any solutions into operation, present the suggestions to the management team. You will need management's buy-in. Ask for their input. Then present the findings to the operators who will be doing the job, and explain the reasons for the changes. Once more, encourage discussion. If any issues are raised, even if you have already dismissed them, take a bit of time and invite the operators to consider them and the consequences of the change. There is no advantage to introducing a new procedure if the operators will not use it. It is also important that you do not introduce an "improvement" that makes matters worse.

Test the solutions to ensure that they work. When it is time to implement them, do not make all the changes at once. In this way, if something does go wrong, the cause will be easier to identify and correct. Introduce the improvements in order of simplicity to implement or prioritize them in terms of the best return on the changes. Monitor the implementation using the plan-do-check-act process. I have included a refresher below.

Plan-Do-Check-Act

What do you expect to happen when the process is implemented? How do you expect to confirm success? What happens if you get a step wrong or the plan is not turning out as you expected? Follow the basic *plan-do-check-act* (PDCA) cycle as shown in Figure 2.2.

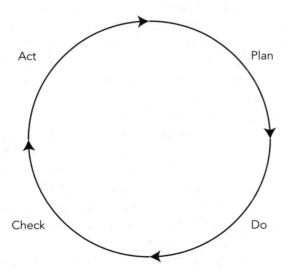

Figure 2.2   The plan-do-check-act (PDCA) cycle.

- ▲ *Plan* each step, and understand the outcome expected at each point.
- ▲ *Do* the task as planned.
- ▲ *Check* that the outcome is as expected. This will involve frequent monitoring. If the outcome is not as expected,
- ▲ *Act*—find out what is going wrong. Then repeat the cycle:
- ▲ *Plan* what you are going to do to put it right and what different outcomes you expect.
- ▲ *Do* make the corrections. Then continue the rest of the cycle, repeating the loop as often as is needed.

As experience in analyzing problems improves, errors will become less likely. If implementation of the solution is monitored, any deviation from the expected plan will be detected *quickly*, and action can be taken immediately to limit any issues. Even experts get things wrong from time to time. The team will be unlikely to know everything; never be afraid to seek advice from outside the team when needed. If the improvements are related to equipment or materials, for example, the vendor is an excellent source for advice. Learn from the vendors' problems. Share experiences.

There are two views on hiding errors. The first is the one I promote: *Don't*. There is always something to learn from any mistake. Why did the team decide on that particular action? Did the team make a wrong assumption? The decision made could have been correct, based on the information available at the time. Perhaps more information was needed. This is one of the reasons that continuous improvement schemes repeat analyses regularly. Turn the mistake into a positive. After all, to progress this far, the managing director and the management team have committed to buy into the whole improvement process because they know that the gains will be well worth the effort.

The second view is less positive. It is influenced by the company culture. Culture is the one characteristic that causes most projects to fail. Owing to the importance of culture, I will refer to it throughout this book.

In this instance, I will consider only one aspect: the "blame" culture. Successful projects need to consider "warts and all." Blame cultures tend to hide the reasons for failures. Indeed, sometimes they even hide the failures—literally. I watched one company actually change from a "no blame" to a "blame" culture over a period of time. One excellent engineer made a mistake; he forgot to close a valve. The gas wiped out the entire process and coated the inside of the machine so thick that the door would not open. No one would admit the mistake for fear of repercussions. This would not have happened as few as six months before.

In a blame culture, a "difficult" boss often will look for someone to blame when anything goes wrong. This discourages people from telling the truth about issues. The same boss also may blame others for her mistakes. Initially, the source can be just one person in the company, but it can permeate through a team and tends to become a culture that is embedded in management. In this situation, the project team will find it more difficult to make changes—in case they go wrong.

To avoid these issues, just follow the rules. Take a look at the decision-making flowchart in Chapter 7. Consider what can go wrong and what can be done to avoid or minimize any impact. Use the brainstorming techniques. Run all recommended improvements past the difficult manager (as well as other managers) in advance to ask her advice and to get her buy-in.

How Do I Analyze the Process?

Whenever anyone thinks about productivity improvements, they immediately jump to the manufacturing process as the area to be fixed, but this is not the only area where you will find problems (or as the politically correct would say, find *opportunities*). Information flow and communication cause almost as many issues—as high as 40 percent. As mentioned earlier while discussing consultants, a good place to start is to talk to the people who do the jobs. Ask them what problems they have that prevent them from doing their jobs "on a day-to-day basis." (Notice how often I repeat this phrase.) You will get answers along the lines of

- ▲ "I ran out of materials."
- ▲ "I was late because I had to go to stores for parts. It took ages to find them."
- ▲ "There are no parts in stores."
- ▲ "The wrong parts are in stores."
- ▲ "The stores people cannot find the materials."
- ▲ "There are bits missing from the kit."
- ▲ "The parts from the supplier don't fit."
- ▲ "I have to wait for instructions."
- ▲ "I don't know what job to do next."
- ▲ "They keep changing the process that is set up before the job is finished."

- ▲ "I can't find the tools I need."
- ▲ "The machine is down for repair."
- ▲ "The instructions are wrong."
- ▲ "I have to keep making changes or reworking parts."
- ▲ "The machine keeps jamming, and I need to free it up."
- ▲ "I missed the Fed-Ex pickup."

The answers you get from the operators will be precise: "I ran out of screws." However, it is not the *specific* issues that you need to fix; you must boil them down to root causes. If there were no screws, find out why. Having none is only the symptom of a problem.

- ▲ Why were there no screws?
- ▲ Does it happen often?
- ▲ What else do you run out of?
- ▲ Is it a supplier issue?
- ▲ Where should the screws be stored?
- ▲ Do we have enough of them?
- ▲ Is there a stock control system?
- ▲ Is the bill of materials correct?
- ▲ Are the instructions wrong?
- ▲ Did we update the diagrams and parts lists when we modified the instructions?
- ▲ Is there a minimum stock holding?
- ▲ Does anyone actually check the stock?
- ▲ Were the screws ordered?
- ▲ Did we check whether the parts had arrived before we started running the process?
- ▲ Is the workstation layout/stores layout suitable for the job?

Some issues are so embedded that they are no longer seen as problems. Do you ever hear statements such as, "It is impossible to estimate that job properly. The customer has no idea what he wants." Or "Be careful with that bit; the roller doesn't fit properly." Spend some time to find out what the issues are. Agnes Pollock introduced me to the *magic wand concept*: "If you had a magic wand (or three wishes), what would you change?" The reason for the magic wand is that it can change anything. Nothing is impossible. It creates the freedom to *want* to fix an issue when everyone says that it is impossible. The magic wand tends to encourage bigger-picture thinking—new premises, a better product, or a culture change. I once spoke to a leather producer about what he wanted. He had previously shown concern about waste in the cutting process. He thought for a bit, while his colleagues gave answers similar to the preceding. Then he said, "I think I would like square cows."

A more technical example involves a company that claimed that "the same yarn" would not run on a range of identical machines. Nothing could be done about it. The problem has been around for years and affected everyone. I could not see how this was possible. If a machine is set to a particular pulling strength and is calibrated properly, then the yarn is more likely to be the issue. I cannot describe the actual issue here, but a bit of investigation showed that one type of yarn was prone to becoming weaker over time. A solution was introduced and the issue eliminated.

A third example involves a company where the production staff "wasted" 10 percent of its day getting its own parts from stores. This was what the staff always did. These were highly skilled engineers and operators. It never occurred to anyone that all the time an engineer spends looking for or getting parts from stores, she is not spending any time using her skill making products. From the perspective of production, the task is non-value-added. At the very least, it is *reduced* value added, which is not an official state in Lean. An unskilled person or materials handler should be used to deliver the parts. I introduced OEE in Chapter 1. If you remember, OEE consists of quality, availability, and performance. If there is no operator at the workstation for 10 percent of the time, the *availability* of that workstation is reduced by the same 10 percent.

If the operators spent their time making product, they would increase production by 10 percent (10 percent of a $1 million turnover is $100,000). And remember, all the labor, heating, lighting, electricity, compressed air, and tools are already paid for. The bulk of the extra product is straight profit. The only factor limiting the benefit is that the company must be able to sell the product. Otherwise, it just becomes expensive inventory, which is one of the Lean "7 wastes." (It is the *I* in TIM WOOD—explained later in this chapter.) Items that cannot be sold end up sitting in the finished-goods store and are simply costing the company money. The beauty of this issue is how easy it could be fixed despite being accepted practice for years.

An Introduction to Process Maps

Process maps are basically large flow diagrams. There are two types of maps that I will discuss here: the *big picture map*, as shown in Figure 2.3, and the basic *process map*, as shown in Figure 2.4. The big picture map reviews the whole process at a company level.

- ▲ How orders are generated—on paper, by phone, or via e-mail
- ▲ How the orders are planned for production
- ▲ How materials are ordered
- ▲ Goods-in issues
- ▲ Store issues
- ▲ Main production stages—not individual process steps
- ▲ Goods out
- ▲ Customer delivery
- ▲ Billing
- ▲ Quality and other features

The process map is more detailed. It is used for directly analyzing a process or a production line. For example, the big picture map might include the process step "Oven," which is a complete process in itself. The process map is more detailed. It would break down the "Oven" process into individual steps: Show how the material is brought to the oven, where it came from, how it is stored at the oven, how the oven is loaded, whether the oven is filled, the temperature-setting procedure, the required limits for the process, how many units are loaded, how long it takes to load, how long it takes to heat, how long the material stays in the oven, how the quality is checked, how many employees are involved, and so on. One step becomes many.

It is common to hear people refer to *brown paper maps*. Owing to the size of maps, mapping is carried out on walls, normally on rolls of paper or sheets that can be taken down when not in use or when visitors come. I have seen maps that run for more than 3 feet (10 meters), all the way around a room. The process steps are written on Post-it Notes, one step to each. Why Post-its? (No, I don't have shares in the company.) They can be moved easily when needed. I guarantee that steps will be missed, often several in a row. It is much easier to add data when the steps to either side can be moved.

I also recommend Post-it Extra Sticky Notes—based entirely on my early experiences. There is nothing worse than laying out a complex chart and coming in the following day to find all the steps lying on the floor. I also like using Post-it Flip Chart Sheets (giant Post-its with a sticky top) as a backdrop in place of paper rolls. I much prefer them to rolls of brown paper. They can be laid out in a row, one sheet at a time. For complex maps, extra rows can be added above and below (like tiles) to allow for parallel production lines or split processes. Using Post-it sheets makes it easier to move a bunch of steps to allow for missing steps to be added.

Notice that the big picture map is a closed loop (Figure 2.3). The customers and suppliers are each shown as one unit for practicality. The process that follows is a tad ideal because it ignores the impact of the "7 wastes," but the map is intended only as a taster. Notice, however, simple as it is, that there are a lot of communications going on to satisfy just one sale.

The process starts when the customer checks his orders against his stock levels. He orders the parts he needs from the sales department, which updates the administrative office about the sale. It is the job of the administrative office to turn the customer order into a bill of materials. Any parts required by manufacturing to fulfill the order also will be purchased by administration from its suppliers. Administration also will ensure that the delivery dates are achievable and contact the production department to inform it that the job is coming. Administration, the goods-in department, and the warehouse discuss the order and are told when to expect the incoming parts. The warehouse confirms existing stock levels and feeds back to administration.

The production department has to schedule the job, confirm the delivery date to administration, and prepare any manufacturing job sheets. It also needs to ensure that the equipment will be set up and working when needed, operators are available, and everything will be ready to begin. This includes communicating with goods-in and the warehouse, which will track the supplier's deliveries and ensure that the goods arrive on time and are of acceptable quality. The warehouse will prepare the materials for each stage of the process and ensure that they are delivered to the operators when they are needed—not before—and so avoid cluttering the production area.

The quality will be checked at appropriate stages of the process, issues flagged to production, and finished goods sent to the store or directly to the customer. The goods-out department, having ensured that production will be completed on time, will confirm delivery dates with the customer and arrange transport. The customer, in turn, reviews his current needs and places another order with sales to flow through the process again. What could be simpler? (By the way, I kept the magic wand!)

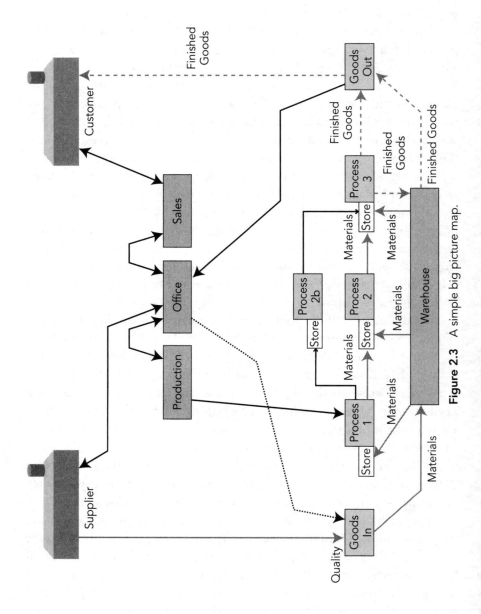

Figure 2.3 A simple big picture map.

In Chapter 3 we will look at the big picture map in detail.

Figure 2.4 is a simple process map that illustrates a standard production line with two options that can be selected by the customer. When we reach a specific point on the main process, depending on the option wanted, the process will follow one of the two parallel processes. At the end of the optional process, we are fed back to the main process.

When creating the map, a color code is used. Any colors can be used, but once chosen, I recommend that the same colors are standardized for every map. This will mean that anyone viewing any map will know what it means. I recommend yellow for process steps because it is easy to see; red for problems because red is always associated with problems, negative numbers, and losses; and blue for possible solutions. Dave Hale use to say that ideas come "out of the blue." I like that—it sticks in the mind. Where specific forms or documents are used, they, too, should be positioned on the map, close to where they are used.

Maps can become very complicated. One advantage of Post-its is that you can split out a complex part of a map simply by removing a section of Post-its and start a separate map on another wall.

The two options are shown as single lines, but the options themselves also can split into parallel tasks. Some very complex maps can resemble train tracks around a busy station.

On completion of the process steps, for both types of map, each yellow Post-it is reviewed in order by the team, and any issues are recorded. If we look at option 1 in Figure 2.4, we can see that steps 2 and 5 have only one red issue. However, in option 2, we can see that steps 1 and 6 both have three red issues.

If we imagine that the process is making a cake, option 2 might be making the high-end version or a birthday cake, for example. The problems might be something like having no icing, the wrong icing, the mix is too thin, the piping attachments are missing, or there is no one able to pipe the wording. We are looking for the problems that affect production. Use the Lean "7 wastes" as a guide.

Quantifying Losses

The red issues can be transferred to a spreadsheet and converted to numbers. This can be lost time, numbers of units scrapped, the cost of extra

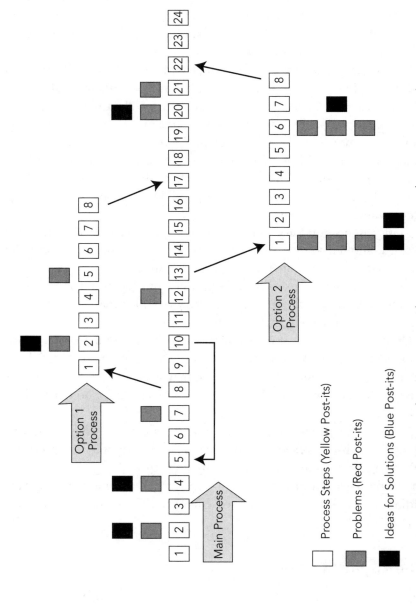

Figure 2.4 A simple process map with one main process and two customer options.

materials, and overtime payments. We want to turn each issue into a figure that can be compared with all other issues. This is one of the reasons money is a good way to estimate the waste. Initially, the operators in a class will not like to think in terms of money. It has to be explained that this is an ideal unit for comparison. Time lost has a dollar value in personnel-hours, wasted electricity and other facilities are easy to convert, and materials also have a cost, as does an express delivery charge. Reconsidering the OEE example, waiting for an operator can have a dollar loss owing to product not being made—as can all the other time delays.

If we sum the estimated costs for each single loss, we then can count how often it happens over a year and calculate a total annual amount. If we repeat this with all the losses, we can quickly identify which are the most costly issues. Remember, it is not the cost of a single event that defines the size of a problem but the number of times it happens.

The "7 Wastes"—With the Cake Icing Examples

The "7 wastes" have been mentioned frequently and will continue to be. This is so because they represent the most common types of issues and they apply to every company. They will be covered in more detail in Chapter 5. I cannot overemphasize how important they are. In addition to *value* and *pull*, they are the most useful features I have found in Lean. Before I learned the TIM WOOD acronym, I had difficulty remembering them. Having said that, I also have difficulty remembering the names of the seven dwarfs.

The "7 wastes" are listed here:

1. *Transporting.* This is any extra cost caused by delivering materials, for example, delivering new icing or piping attachments to the workstation. Both require the operator to wait, so we also have a wages cost and a production loss.
2. *Inventory.* No icing in stock can create a long wait as the goods are ordered and delivered. If production has stopped, we have a productivity loss as well as the cost of operators waiting.

    To reduce the waiting loss, the product being made could be changed to another process. This creates a different loss owing to the time taken to set up the line for the new product. We also have lost

production for the time taken to return to the original process. We also could have late deliveries to customers. These could involve overtime rates or express deliveries.
3. *Movement.* Movement is what we know as ergonomics and factory layout.
4. *Waiting.* All forms of waiting cost time, and time costs in wages and lost production. I mentioned a waiting loss in step 2, but I intend to repeat it here: If there is icing in stores, the icing has a delivery cost just as in defects. This will be a relatively low cost. No icing in stock, however, can be a longer wait increased by the time taken to set up the line to make something else.

   Remember to include the time taken by the operators to change the line as lost wages and lost product. The mix being too thin requires waiting for a new batch to be made.

   Missing piping attachments requires someone to find and/or clean them. No one able to pipe the wording means having to wait for the operator.

   Waiting is a major problem.
5. *Overproduction.* This is simply making too much product. The goal is to make only what you know can be sold. In the case of cakes, overproduction means scrap or selling off to employees at a cheap rate.
6. *Overprocessing.* This is having too many steps in the process, some of them unnecessary; for example, applying two layers of the same icing and not just one.
7. *Defects.* The mix is too thin. This is a defect in a part of the process. If the mix can be salvaged, we have a rework cost that takes time in personnel-hours and some extra material costs. If it has to be scrapped, we have a material cost for the scrap and the time of an operator to make the new batch. We also need someone to deliver it to the workstation. Plus, the original operator has to wait for the material to arrive, so nothing is produced until then, which is a reduction in throughput.

   It will not take long to get a feel for these issues. A simple spreadsheet could be designed to ask the questions. In this way, there is no need to remember all the possible losses. Besides, we are only making an estimate to get a feel for what the losses are and, consequently, what the gains will be. The estimates can be corrected easily as new ideas surface. The key point to remember is never to overestimate the savings. If possible, include someone from finance on the team. If estimates are excessive, the credibility of the team will be lost. It is better to underestimate and be pleasantly surprised.

CHAPTER 3

The Big Picture Map

Before starting the map, consider a short refresher for the team as to why mapping is needed. You have a cross-functional team of people that you hope will be willing to improve the way things are being done. This is the group that will make or break the project. Are they willing participants? If they are, your success rate has just shot upward at an exponential rate. If they are not, the picture has the potential to be less successful.

There are two things you can do here: The first is to have a team that you know will participate; the second is to plan to spend more time working on motivation and mentoring. It will take a great amount of coaching and discussion to ensure that no one feels threatened by any of the issues the team will discover. The task is to find ways to make improvements—or opportunities, as we say in politically correct talk—*not* to find people to blame.

Before starting the map, I like to take a bit of time to paint the background—to remind people of the need for the exercise and the issues that led us to this point. Chapter 2 covered the problem-solving technique as an initial means to get a general feel for the production issues. I can recommend another two ways, the first of which has been mentioned before because I use it so often.

I normally have one day to make an initial *snapshot* assessment of a company, so the method I use most often is simply to talk to the people who work there. When talking to senior management, my meetings are more formal. I use notes to help me remember the questions. If I did not use the prompts, I might forget what to ask. There is always the danger of a conversation getting bogged down on a specific topic. The list keeps the questioner (me!) on track. For the production staff, my objective is to find out what *they* think stops them from doing their jobs. The questions are more general, but the answers will be precise and detailed, inviting further analysis. For management, one question I like to ask is what keeps them awake at night.

The second method is to carry out a strengths, weaknesses, opportunities, and threats (SWOT) analysis. The SWOT analysis and the big picture map are strategy tools best carried out by managers and section heads. Both look for the bigger issues, not specific process detail. These issues may include cash flow, supplier or customer issues, getting sales, design issues, not meeting orders, overcrowded stores, unreliable processes, equipment upgrading, or factory layout.

The SWOT Diagram

The best improvement techniques work because they involve a degree of formality. Unfortunately, I believe that this is also why they are underused. I like them because the formality tends to make the issues less personal. SWOT analysis, like brainstorming, should not be reserved only for training courses. SWOT analysis is an "ideas" tool. It will quickly highlight some good opportunities for improvement and some reasons for doing so. In a SWOT diagram (Figure 3.1), we consider four key aspects of the company: strengths, weaknesses, opportunities, and threats. It is always best to work in teams and use brainstorming to gather the key points together.

1. *Strengths.* What is the company good at? Why do your customers buy from you? Every company has its own set of skills that it uses to produce its products. Don't confuse skill and knowledge: They are not the same. Knowing how to do something is not necessarily the same as being able to do it.

    As examples of strength, sales may be a strong area or the quality of the product may be excellent. When listing the points, listen to what is said by team members. "Sales are excellent—except when we make the . . ." or "We tend to rely on previous customers . . ." or "We can get some really good design ideas—if we could only make them work" or "We have great ideas—but they end up costing too much to make or no one wants to buy them." Count the good half of the comment as one of the "Strengths," but also consider the negative half. Find out why the strengths do not always work, and list the answers under "Weaknesses."

Strengths	Weaknesses
Sales	Can take too long to complete a design
Design team	Can cost more than competiton
Build quality of product	Too much money tied up in stores
Good control of cash flow	High rework costs
Factory and equipment	70 percent on-time delivery
Wind power units for homes	Lower prices from low-cost economies
High-power vacuum systems	Bank interest rates
European agents	Internet access to foreign competition
Investigate portable solar power generation	Sourcing raw materials
Providing technical support in the Far East	
Opportunities	**Threats**

Figure 3.1 The SWOT diagram.

I add the next point knowing that you will agree with it and yet also knowing that there are companies where it is frowned on. You must be honest in the analysis. Don't just agree with your teammates.

2. *Weaknesses.* This is the other side of the strength coin. What do you know you are not happy with or think might not be up to scratch? What can you *not* do to a satisfactory standard? For example, do some tasks take a lot of rework to get them right? Do you create a lot of scrap or only discover mistakes at the end of the process? Is there a job only one person can do, and you have major problems when she is off? How often do you hear, "The delivery probably will be late"? Specific issues are acceptable: root causes are way better.

3. *Opportunities.* What is happening in the marketplace now? What can you see developing in the future? Do you have extra capacity you can fill? What do your competitors make that you can do better—or as good as? What are the salespeople asked for that is not a current product? Is there anything you can make to get into this market? Is there an opening

in the new, developing areas where can use your skill set? Is there a new skill that the company needs to learn that will open up options?

If you currently make electric fans, can you maybe make small wind turbines for generating electricity for homes? If you weave textiles, is there a new material or product in construction, disaster relief, photographic printing, medicine, or another field?

4. *Threats.* This is the opposite of opportunities. Consider any current or potential threats that are facing the company. Any cash-flow issues? Perhaps you never introduce new products? Is your production equipment past its use by date, unreliable, or slow? Is your factory too small or too big? Are foreign companies competing in local markets? If your company prints invoices or advertising posters, are the developments in electronic communications, computerized billing, or the Internet likely to affect your business? Is there a threat of cheaper products being made in the developing markets? Are there any new laws that will increase your energy or waste costs?

Brainstorm for ideas, and discuss all the issues at the end of the process. Creating a simple SWOT diagram can take as little as an hour, but a more detailed one can take a morning. Try to get input from all the managers and key employees. Consider using Post-it Notes, one for each idea. They make the analysis easier. Post-its can be regrouped easily into similar themes, making root causes easier to find. If there are lots of ideas, the SWOT diagram does not have to be contained on a single sheet of paper. Use the individual pages of flip charts for each of the four sections.

The Big Picture Map versus the Process Map

The two maps being discussed here are the ones that I find are used most frequently in Lean diagnosis. There is no one way to create these maps. This is a feature I like. I think that maps should be created to provide what you need. While researching this book, I did a Web search. Google returned 107,000 results in 0.19 second! There are lots of maps out there. In some cases, the search for "big picture map" simply returned references to a process map or a value-stream map. This is so because the maps evolve over time. Each time a map is used by a different team, the team will introduce changes from the way the process was taught. I tend to drift toward simplicity. For me, the goal is not perfect maps but perfect solutions.

Now we prepare for the *big picture map*. This is a high-level flow diagram. (It has always confused me that *high level* means less detail!) Anyway, just to get you into my habit of bending the rules to suit your own needs, feel free to use a bit more detail in any parts where you feel that more information is necessary. The chart exists to help you. I would always be as flexible as I need to be. I have even used a "blended" big picture map that had an essential process woven into it. It took the form of a second map, to the right of the "customer" box. It looked a bit like the handle on a frying pan, with the loop of the big picture map being the pot. It introduced key details that the client felt had to be considered.

If you want to increase the detail on too many points, perhaps you need to use a *process map*. This is the most common map, and it is used to find the maximum information. It can be applied to a whole company, but it will be a huge map. One map I created previously went all the way round a conference room. Process maps tend to be best when they are applied to specific areas, but they will always end up bigger than planned because they naturally spill into other areas as they develop. Being much more detailed, such a map takes longer to put together, but it will highlight many more opportunities.

The background information gathered by talking to employees provides a spectrum of current issues and problems. The list will be wide ranging. Most will be practical productivity issues, more suitable for a process map. Other issues will be those that individuals find irritating, are pet peeves, or are previously ignored suggestions. Don't ignore them. All will be useful. The issues might be *symptoms* of bigger problems that will lead you to the root cause. Fixing a symptom will not resolve a problem. Fixing a root cause will. If the same problem has appeared more than once despite action taken to resolve it, it is very likely that the root cause has not been discovered. The goal of the analysis is to find and resolve the root cause for *each* of the issues. If the symptoms are linked, the same solution may solve a number of problems.

Production Logs

One downside of analysis by discussion is that people tend to have a limited memory of issues. Employees think mainly about their current problems and often forget those more than a couple of months old. Use the team to analyze the production downtime log—if you have one.

If you don't record all issues that interrupt production, I recommend that you introduce a log immediately. Materials resource planning (MRP) and enterprise resource planning (ERP) systems normally have a log included, but a book is still a good place to start. The log should record all the reasons for downtime—quality failures, equipment breakdowns, maintenance periods, maintenance overrunning, operator shortages, equipment adjustments (such as setting a machine to run slower than it should), temporary fixes to equipment or processes, setup problems, product changeover losses, problems with the process, material shortages, unplanned changes in the production schedule, and problems with instructions. Don't forget any incorrect repairs while trying to find the real fault. In short, record everything, and set up a system for analyzing the issues.

You need to consider the impact of the "7 wastes." If you analyze a machine breakdown that fails at 7 a.m. and returns to production at 11 a.m., the total downtime is four hours. But what if the repair took only 30 minutes, and the rest of the time was spent waiting for an engineer or for tools or parts? Sometimes companies wait days for a vendor engineer and weeks for parts. You need to understand all the relevant information about lost production time. What if you then estimated the annual losses as a means of standardizing the costs and discovered that you lost two or three times more money each year waiting for engineers than it would cost to hire one of your own or to hire a second engineer? You also might discover the advantages of investing in carrying some key parts in stores.

A log is essential—but it must be used. It is not a diary to be read by the next maintenance manager. Don't just record information if you don't intend to act on it. The log even will help in fault finding. I once discovered the solution to a major fault in a customer's site simply by referring to the log. Each time the unit failed, it took days to get back online—production losses and personnel costs were huge. The company's board of directors was so angry that they had summoned the managers of my company to explain the situation. I only found out there was an issue when I was told that my bosses were coming. I found the problem in a few hours simply by looking back at what had been done to the machine just before the problem first appeared. Fortune would have it that one of the company's own engineers had replaced a component with a different part. (He knew it was a different part but did not consider the implications of the shape change.) This had the effect of reducing a mechanical movement intermittently. Change the fitting, calibrate, and job done. The cost of lost production to the point of the fix was in the hundreds of thousands of dollars.

Employee Input

It is good practice to regularly discuss improvement suggestions with the employees who do the work. If an employee has suggested an improvement, let him or her know that the issue is being considered. If it is not a good idea, is too expensive, or is difficult to implement, explain to the employee who made the suggestion why the idea has not been implemented. The employee may rethink and come up with a better idea. Suggestions that are dismissed with no explanation can prevent employees from making suggestions in the future. This can become a barrier when those employees start to work on continuous improvement teams.

The Big Picture Map

A factory or process can be analyzed in much the same way as any other issue. In place of technical systems diagrams, PowerPoint flowcharts, or engineering drawings, Lean will use the *big picture map*. This map is a systematic way of visualizing all the main functions within a company *and* the key connections between the departments. It starts and ends with the customer. It is designed to be a closed loop from customer order to parts delivered and considers all the essential support stages in between.

In fault or problem diagnosis, it is common to subdivide an operation into *functional* areas and to check the operation of each of the areas in sequence. This helps you to find out what works, to see how the areas are connected to one another, and by elimination, to identify what doesn't work. Sometimes the root problem turns out not to be the main process as such but how two subprocesses (or departments) interact. Analyses can be targeted in areas where a problem is *thought* to exist. My preference is to spend a bit more time in the analysis stage, cover a wider area, be a bit more methodical, and use logic to develop an overall strategy. Instinct can be a help, but it also can be misleading.

The big picture map reviews the operational departments of the whole factory. It sounds complicated, but don't be wary to try out techniques such as this. Even if you get bits wrong or miss something, you can't fail. Every map will teach you about the issues you have and show you something that needs to be improved. The more detailed *process map* and what I have called the *capacity map* will be discussed in Chapters 4 and 5, respectively. However, for best results of a first-time diagnosis, the big picture map is the place to start. Then it can be followed up by a process map and/or a capacity map of the areas where issues were seen.

A big picture map mirrors a company's functions and so consists of *multiple* systems. It includes all the key processes, materials, and information flows. To get the best outcome, the map should be created by the management team, with added information being sought as the need arises. A visual representation allows the team to "see" what happens in *and* between the major production steps, the different departments, the customers, and the suppliers. If compared with a geographic map, the big picture map would look at towns and cities, not streets.

I have seen a few versions of this map. The best, by Agnes Pollock, used cut-out drawings to represent factories, trucks for deliveries, data boxes for explaining process details, boxes for general information on personnel and shifts, and different types of arrows for modes of communication (e.g., e-mail, phone, letters, and fax). A *Q* represents a quality inspection point, and an *I* in a triangle represents a place where inventory is stored, even if only a temporary store. The graphics took Agnes a lot of time to prepare, although there probably is a store somewhere where they can be bought.

As a trainer, you may want to impress the management team and make the map look more professional. There is often a need to consider the audience and what its members like. To save you drawing the icons, I recommend a quick search of the Web to find downloadable symbols. Some sites will have different symbols that you can use if you prefer. There are also software packages that can be used to draw the maps. Such packages make nice maps, but even if the monitor image is projected at a high resolution on a very large screen, I still recommend the manual method. I like a simple Post-it map with hand drawings or arrows with the word *pull* or *push* or *I* or *Q*. This is so because I like to promote improvements as not needing special, expensive tools—just simple processes with $15 worth of sticky labels and some brainpower. In addition, I find that "handling" the data brings the team together.

What Do We Need to Know to Make a Useful Map?

Production is not the source of all losses. Of the 40 percent generated in the support departments, communications and information are huge contributors. In any company, information moves in all directions like the ball in a pinball machine—bouncing to and from multiple departments, customers, and suppliers. It arrives by phone, text, e-mail, letter, fax, Post-its, and scribbled notes. It comes from people, can be processed more than once, and occasionally gets lost and doesn't come at all. It goes to the wrong person or sits on an empty desk for weeks while the person is on vacation. When you consider all the processes used in a factory on a daily basis, how does anyone know what to do?

To put a basic map together, there are at least seven distinct inputs to consider. These are all incorporated in the map. They could be considered in a couple of different orders. I like the following list because the sequence of the steps is defined for practical reasons.

Before proceeding, notice that I said a *basic* map. The map is your process. In the original maps, we look for issues to fix using the "7 wastes." The issues become the red Post-its. But having the entire process laid out before you enables you to look for anything. What if, for example, you wanted to carry out a risk analysis of the various departments (or areas or process steps)? When you review the process steps, you can ask, "What can go wrong here?"

If you are making furniture, for example, using the wrong veneer or stain could eat up a lot of time to put right. If you are making precision or complex glass items, what if one is dropped? At the start of the process, this would be less of an issue than at the final stage. If you were making medicines, what is the risk if the refrigerated storage container or a key piece of equipment fails? What if the autoclave is contaminated from a previous breakage or an incoming gas line? We could use the analysis to review risks or the cost of scrap or rework as the job moves through the process. What about operator cover? Is there an area where you would be in trouble if the operator was ill? In short, you can analyze the map for anything *you* need—or everything you need.

When I look at the layout of a factory, all the different areas, equipment, side rooms, shelves and stores, and desks and offices, I have no idea how they relate to each other. A guided tour does not give enough detail, and I know that I will not remember all of it. Besides, the areas are highlighted by the guide as we pass them, not in the order of production. Indeed, some of the production departments can be located in different buildings or special machining areas or clean rooms. So how can I easily define the actual process path? I will explain the most common technique. If the logic of the steps is followed, it is difficult to get the map wrong.

What do you know that passes through all the production departments? Raw materials and product—they must pass through all the departments where they need to have work done to them. (Remember to consider different products, too, if they use different tools.) If you follow the material from the suppliers into the company and through all the various stages and departments, you end up with a representation of a physical layout.

What should be the second step, then, a rework line? It does have a physical presence. No. Information is more important because it originates in several office-based departments and accounts for nearly half the issues—sales, administration, purchasing, design, quality assurance (QA), and production. Information is essential because it ensures that each department gets what it needs—when are parts arriving and what to do with them and by when.

It is possible to change the sequence, but the order here is fairly standard. Steps 1 and 2 are essential: They define the factory flow. As discussed earlier, rework, for example, now may make a better step 3, but a vast number of companies don't recognize rework. It is an actual process step (or loop), albeit an undesirable one, but we don't know where the issues arise yet. Interestingly, the more essential a rework loop, the worse is the performance of the process. So what do we need next? Production data.

The remaining steps look for specific Lean functionality: How long does it take to make the parts or pass through the process? Do we have bottlenecks or online storage points? Where do we check for quality? Only now that we have quality checks can we know what has to be scrapped or reworked. So the final stage is good for the rework lines.

In summary, the mapping steps I will follow are:

1. Materials
2. Information and communications
3. Production statistics and data
4. Lead times for each stage
5. Inventory storage, stocking points, and quantities held
6. Quality and inspection points
7. Scrap and rework loops

(If we were analyzing for risk, the quality and inspection points should mark the steps where inspection minimizes the risks. Remember, the map looks at the process as it is now. The future state map, shown in Figure 3.7, is where improved positions would be listed.)

If we follow the preceding sequence, assembling the processes in layers and sticking to the recommended map format, we will develop a good overall layout. This also will make it less likely that anything will be missed. Figure 3.2 shows a basic map. For practicality, not all the data boxes are shown, and they are overscale. A proper map may be 9 to 12 feet (3 to 4 meters) long and as high as one or two rolls of brown paper or wallpaper. If you want to see a more detailed map, many are available on the Web. The illustrations here are intended to show the process.

As the team starts to build the map, all the issues team members know about will pop to the surface like bubbles in water. "The deliveries turn up when it suits them, and we all have to drop what we are doing and start unloading" or "The trucks have to wait for ages and block the yard (or the street or the local car park)." This is not anything to worry about. Since it happens to everyone, I keep notes in my own "parking lot."

The Parking Lot

There was a time when I pushed for perfection from the outset. Now I've mellowed. Now I don't believe that you need to be an expert in the mapping process or even in Lean—just be able to use them. Skill will grow with experience. Use a "parking lot" to record any issue or information that you think is a problem or simply don't know what to do with. Take a page of flip chart paper, stick it on the wall, and head it as "Parking Lot." Record the information as a series of bullet points or as individual points on Post-its. The parking lot can be subdivided into different headings. Alternatively, if

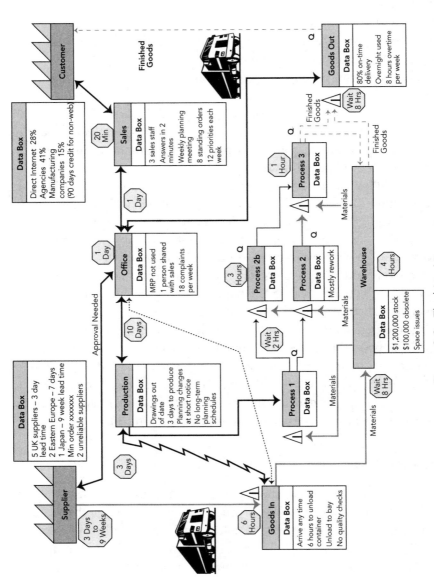

Figure 3.2 The big picture map.

there are too many issues for one page, you can have more than one lot with a page for each group. My preference is to use Post-its.

Later, during the analysis stage, when the team revisits the flow to find the main issues, team members can review the parking lot, analyze the issues, look for common themes, and group the Post-its into categories. Just as in brainstorming, the only no-no is *don't* try to find solutions at this stage.

One point to remember: Make certain that you understand the problem when you write it down. Use simple, plain language to ensure that when it is read later, the meaning is clear. A team I once worked with listed a problem on a Post-it. We had estimated a cost to the company at $120,000 (£75,000) a year. When we came back to it later, we could not remember what the note referred to!

Material Flow

This is the first *layer* of the map to track. By necessity, it defines the departments and the paths that the raw material moves through from the external supplier to the company, through the warehouse and production to dispatch, and then on to the customer. Following the material flow helps the team to appreciate:

1. The physical layout of the factory
2. All the different departments and what they do
3. Where inventory (storage) is located
4. Where there is limited space and even where materials block passageways

The map even can highlight where the rubbish is dumped (Figure 3.3). Think about the processes used. For example, does the stock system consider everything needed to carry out the manufacturing process from major assemblies to nuts, washers, and labels? How often do you run out of parts? If one critical component can stop you from making your product, you need to know about it and make sure that it does not get the chance to bite you in the butt.

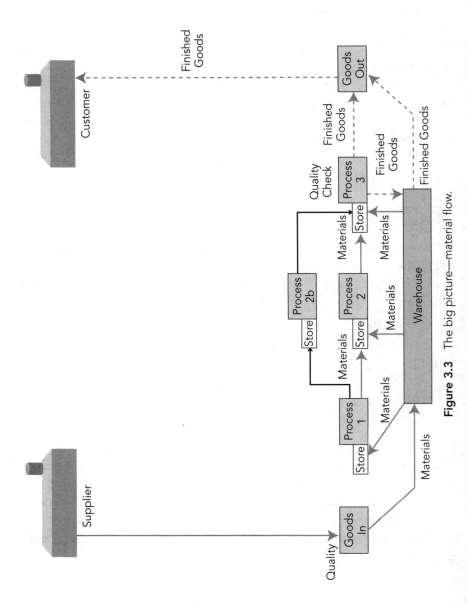

Figure 3.3 The big picture—material flow.

Kanban

Kanban is not a part of the mapping process but is an important Lean technique.

Eventually, every inventory item should be risk assessed to find out the impact it will have if it runs out of stock or if the supplier goes out of business. Remember, size isn't everything. Just because screws and washers are small and easy to get, don't forget about the lost production time it takes to actually get them. What happens if you run out at night or on the weekend? Don't assume that employees will warn you when parts are running out. A fail-safe reorder system—such as *Kanban*—is recommended to protect yourself and to avoid shortages. Employees will notice that parts are missing only when they go to get them and find that there are none left. Even purchasing will think that it is safe as long as the part has been ordered. Purchasing rarely checks to confirm whether a part actually has arrived as planned.

To illustrate this point, I will tell you of one factory that ran out of washers: Production stopped dead. The total loss was more than $80,000 (£50,000) because it happened on a weekend. Neither the operators nor the automatic tools could assemble the units without washers. A simple two-box *Kanban* storage system would have avoided this.

Kanban uses a visual signal to act as a physical reminder to reorder. It can be as simple as:

- A laminated sheet of paper saying "Reorder" partway down a pile of boxes. When the sheet becomes visible, it is taken to the person who orders the parts.
- A card in a tray that says "Reorder quantity = 2." This is good for small numbers of parts.
- Red paint on the floor or on a shelf that becomes visible when there are no parts on top of it. When the trigger part is removed, seeing the red paint signals that an order is needed.
- A minimum level marked on the side of a shelf.
- A piece of tape across a box that is broken when someone puts his or her hand in the box to remove the item. The broken tape signals that a new part is needed.
- A two-box system.

The two-box system is exactly as it sounds. There are two containers of the same parts nose to tail in the stores. When the first box is empty, the empty box is the flag used to trigger the reordering. The second container must hold enough stock to last until the new parts arrive. Each box should be labeled with the part description and the minimum order quantity. But remember to adjust the minimum order quantities in line with product demand.

If you have a computerized material resources planning (MRP) system, it is essential that regular stock checks are made. The people who enter the data or take the parts out of stores cannot always be relied on to record the proper information. Indeed, if an MRP system does not track the real factory process, a computer can cause chaos. I have worked in companies that believe that the best stock control system is a padlock on the door.

Materials—Step 1

Prepare a base for the map. Use brown paper or white wallpaper that can be written on. It should be about 10 feet (3 meters) in length. Tape it to a wall so that the top of the paper can be reached easily.

As shown in Figure 3.3, draw two factories on yellow Post-its. One will represent your customer(s); the other will be your supplier(s). Place the supplier(s) on the top left-hand side and the customer(s) on the top right-hand side. You probably will have multiple suppliers and customers, but to avoid clutter, one factory will do as a representation. (A "data box" with a list of suppliers or customers will be added later.) The map of the process will flow from left to right, starting with the materials being delivered to stores. Notice that the drawings need only be simple representations.

Occasionally, a team will insist on doing things its own way. Since Lean encourages empowerment, I feel that I have to be flexible. In one company with a unique product, the layout the team wanted to use was more linear, like a process map. I could not persuade team members to use the proper format. The map worked, but we had some very complicated communication and flow lines, some well over 10 feet (3 meters) long. Whoever drew the layout as it is in Figure 3.2 was a genius. It is the best I have ever used.

To develop our map, if we follow the layout in Figure 3.3, we will have the suppliers on the top left-hand side of the map. The goods will be sent from the supplier and will arrive at the factory. They most probably will arrive at the factory gates by truck, but delivery also may involve a ship, aircraft, or customs if parts are imported. The company may not have a proper goods-in department, but it will have some kind of *booking-in* function. At this point, we are interested in the flow of the materials—where they go.

The questions to consider are, When do the deliveries arrive? How often and how many items at a time? and Where do the goods go when they arrive at the factory? For the layout, the first pass is what actually happens. Remember the parking lot for what should happen.

1. Do the drivers report to security?
2. Do they stop at a weigh bridge?
3. Do they go direct to the loading bay?
4. Do the trucks have to wait to be unloaded?
5. Are the goods unpacked and stored immediately or unpacked outside and brought into stores later?
6. Are the products weighed in the factory?

The following is data-box information. This is out of sequence but always seems to surface as the team discusses this area.

7. How long should it take to unload the goods?
8. Are the deliveries palletized? (See "Lean Note" below.)
9. Irrespective of the way the goods are loaded on the trucks, how many people should be involved in the unloading and sorting?
10. Does anyone need to be taken off-line to help with unloading?

> **Lean Note:** This note links to questions 8, 9, and 10. When the data boxes are added, we want to be able to get an idea of how much operations actually cost. As a process, Lean has objectives that are designed to control costs. Some companies offer to ship goods loose. If the goods are not loaded onto pallets, more items can be packed into a container or truck. This will reduce the cost of shipping and ultimately the unit price for the items. To be sure that this is a real savings, the total cost has to be evaluated and expanded to include all the consequences of the delivery. The most obvious cost is a need for extra personnel. Pallets can be unloaded with one person and a forklift. Unpalletized goods require personnel—lots of people. Sometimes they must come

from production or other areas. It also takes several hours to sort the goods and to confirm that the delivery is correct.

From a Lean perspective, buying in bulk is rarely seen as being a positive. Lean promotes buying only what is needed to sell in the short term. This is not always simple because the quantity to be ordered must consider the *lead time* (the time it takes from ordering the goods to delivery). If a container is being shipped from the Far East, delivery can take weeks. This introduces a conflict for Lean because short-term ordering may not be practical. The cost comparison involves more than just the unit cost of the goods. By buying so much more of an item than you need, other problems can be created.

- ▲ It necessitates spending more money up front, which can affect cash flow and bank interest.
- ▲ It takes longer to get a return on the investment.
- ▲ The cost of the extra labor has to be evaluated.
- ▲ Lost production for unloading—if any—needs to be evaluated.
- ▲ Wrong parts being delivered can be a problem to put right.
- ▲ If your warehouse becomes disorganized and cluttered by the large quantities, finding parts for production becomes an issue. Confirm whether manufacturing is delayed, causing late deliveries or more expensive shipping.
- ▲ The extra material takes up valuable space in the warehouse, with all the associated storage and handling costs.

Continuing with the map, what happens to the materials when they leave the warehouse? Do they go directly to production? What line do they go to? Are they delivered in batches to a rack or location next to the first production step? If so, it can be added to the map, even though inventory details are added later. Do the materials pass to a quality check? Are they double handled at any points? (Add to the parking lot.) Where to next? Follow the process all the way to the customer.

Is it possible to represent the production line as one unit, although it might be more practical to divide it into major functions—for example, paint spraying, bottling, or final test. List each step/stage on a yellow Post-it, *one* per sheet. Where production lines are very different, you may prefer to split processes. Provided that there are not too many, I prefer separate lines for each process because this suits the way I think, but it is not necessary. If the line splits into two processes, represent them on the map as two parallel lines of Post-its, one above the other. Think of the lines as water pipes splitting to go to different places. Follow the pipes through manufacturing to quality to packing, goods out, and the customer.

The information on personnel-hours, unloading times, handling capacities, equipment availability, numbers of forklifts and pallet trolleys, available space, and cost of storage will be added later with other key information in a data box. Similar data boxes will be used for production details—units made, number of operators, and so on.

What if you can't remember the process? Go and investigate. I advise the management team to walk the line. As a facilitator, I like to start blind to remind the team how much they actually know about the process. Perhaps a good compromise is a 20- to 30-minute start to establish what is known and then a tour. It is interesting to discover whether what we think happens really does happen. Visit the line as a team. The managers from purchasing or sales may never have seen the entire process before. If details are needed at later stages, get someone to gather the data. Find out firsthand how easy it is to actually get the information.

In Figure 3.2, the goods are stored in a main warehouse with a series of smaller stores ("supermarkets") located at each process step. When the data are included, we will know how much stock is held at each location. For example, the supermarket might hold enough capacity to run the process for five days. This is not a problem if the quantity is small, but what if five days means 100 window frames or 50,000 empty bottles. This will take up a lot of space. This information will be added in step 5.

Information Flow—Step 2

To quote from *Blazing Saddles*, there will be "a plethora" of information issues. None of the material flows, production plans, or product deliveries can happen without information. So who talks to whom? There might be fewer communication lines than you would hope.

In the following example, there were only two communication breakdowns. The first was no contact between goods-in and the supplier. The second was no communication between production planning and goods-in—until it all went bad!

The company was scheduled to make cream of tomato soup. The production operators spent four hours cleaning the lines and changing over the equipment, preparing the empty cans and labels. Everything was ready to start the run when it was discovered that the company had a problem—the cream was not in the factory. Chaos ensued. Phone calls followed discussions and more calls. It should arrive soon. It didn't. An emergency quick changeover to a non-cream-based vegetable soup took another four hours. Then, just as the line was ready to start, the cream arrived. With cream, it is a use it or lose it scenario—and there was a lot of cream to lose. Another changeover took the production line back to where it was in the morning, the only difference now was a complete shift of lost production and operators doing a whole lot of unnecessary work. The situation could have been avoided if the information had been used properly and the cream was confirmed to be in stock before the first changeover.

Lean promotes visual displays. In this case, all we needed was a *changeover board* (Figure 3.4). These are commonly promoted in the single-minute exchange of die (SMED) approach. Such a board plans the production changes in advance and ensures that parts are available before the changeover starts. If materials are missing, you have time to either confirm delivery or change product. Having specific people responsible ensures continuity of information.

To start tracking the information flow, we begin at the customer. How do customers get their orders to us? Do they deal with sales representatives or agents or directly with the company sales department? By following the information flow, you will discover that you need a few more departments to be added to your diagram. In our case, we add sales, an office (to represent general administration), and production. However, there can be other departments, for example, maintenance, design, QA, and administration. The book is limited by space.

As with the materials, if you follow the flow, you will identify all the departments and all the links. All you need to do is find out who talks to whom:

1. *Who does the sales department talk to?* The customer can phone sales directly but also can e-mail, fax, or buy online. Does the sales department place orders directly with the supplier or work though a design or planning department?

Date	Planned C/O Time	Line	Product	Kit Required and Checked	Ready to Start C/O	Team Responsible for Prep	C/O Complete (Time)	Comments	Team Responsible for Changeover
03 February 2010	6:00	1	Cream of Tomato	A	Yes	Shift A		No cream	Shift A
				Specification	Yes				
				Materials	No				
03 February 2010	18:00	1	Vegetable	B	Yes	Shift A			Shift B
				Specification					
				Materials	No				
04 February 2010	2:00	2	Bean	C	Yes	Shift A		Label Due 1-Jan	Shift A
				Specification	No				
				Materials					
04 February 2010	12:00	1	Cream of Tomato	A		Shift B			Shift B
				Specification					
				Materials	No				

Figure 3.4 Changeover board.

2. *Does production talk directly to contracts, maintenance, design, goods-in, or the suppliers?* Does production have a direct link to the customer to verify specifications and deliveries?
3. *How many groups have the authority to place orders?* Stores, administration, production, or maintenance? Do they order in isolation from each other?
4. *Does goods-in keep in touch with the suppliers to get advanced information on incoming or late deliveries?* Does the department pass information to planning on any issues? Does the planning department schedule deliveries to the production lines? What process is followed if the lines are running late and the run will not end on time?
5. *Do the people who place materials orders to the suppliers liaise with goods-in as to when and what to expect?* Do they update production? Will goods-in confirm delivery schedules?
6. *How does the warehouse liaise with the production lines?* What happens when warehouse personnel cannot find materials?
7. *Where do the operators get their instructions?*
8. *Do the operators know the next few jobs that they will be doing, or do they just appear after the previous run has ended?*

All the important communication lines should be included. When you draw the arrows, you can mark the frequency of contacts on them if that information is important. For example, does goods-in talk to production daily, before every shift, or only when there are problems? Separate electronic and paper communications. Electronic communications are represented by a lightning-strike arrow; otherwise, use a straight arrow. See Figure 3.5.

Data Flow and Data Boxes—Step 3

We now have the two main flows, but we have only paths, not details. We need to know when goods-in gets deliveries, how much, and so on. The data boxes should contain the information best suited to the related process. This is normally productivity information—equipment reliability, units produced per hour, number of employees on each production line, how long it takes to make the product, quality information (yield), and my favorite, equipment *availability*—very useful for bottleneck equipment and maximizing capacity.

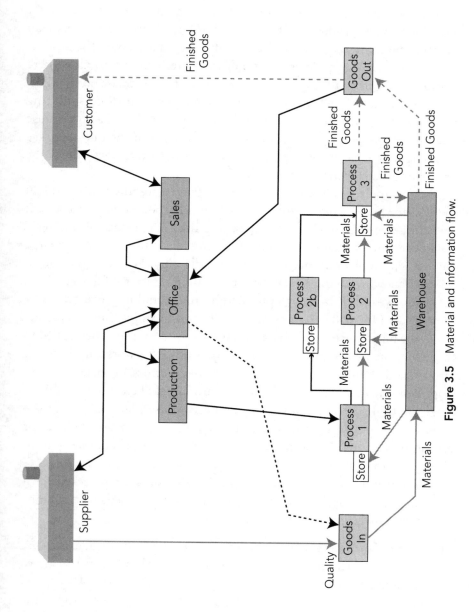

Figure 3.5 Material and information flow.

What Is Availability?

In any given day or week, how much time is the production equipment available to make product *and* how much is it actually used? In Lean, we do not run a machine just to maximize utilization and make units: The product must be able to be sold. However, finding extra production capability where sales are guaranteed adds value to the process and will increase profitability.

Consider a production line where all the equipment stops producing at breaks. It is not a problem on a tool running under its capacity because the loss can be recovered, but it is a serious problem if the unproductive equipment is a bottleneck machine because the bottleneck limits the production rate for the rest of the process.

For example, assume that workers get two 10-minute breaks and 40 minutes for lunch. This is a total of one hour of production time lost each shift. In a five-shift week of 40 hours, you lose 12.5 percent of the week. Using the data in Figure 3.6, where the company is making 68 units in a day, 12.5 percent more machine time (availability) would allow you to produce an extra 8.5 units each day, or 42.5 more units in a full week. This is almost two-thirds of a day's production. You would improve profits by $120,000 to $160,000 (£75,000 to £100,000) by this simple improvement.

If the bigger picture is considered, the same example can be applied to the people on production lines. If someone is not at his workstation for the same period of time, the same loss in productivity occurs. When an engineer stops working to go to stores, nothing is produced. She is not available to make anything. The lesson here is the need to use material handlers—people who act like waiters and deliver components to the engineer or operators. The less time a skilled engineer or operator spends away from his or her workstation, the more product is produced.

To return to thinking about the productivity data, I have produced a list of questions. Not all these data will be included in the data box, but they will be extremely useful for the process map.

1. *What is the frequency of deliveries to the factory?* Where possible, we don't want all the deliveries arriving at one time. If all the deliveries arrive late in the day, for example, we might need overtime to book the parts in. Smaller companies often simply have to accept their suppliers' conditions. Good suppliers will try to help, and larger companies can have some say in delivery schedules.

Process 1	Actual	Best Possible
Production units per shift (average)	68	100
Productivity per person (units/shift)	17	25
Number of shifts	2	
Cycle time for Process 1	60 mins	51 mins
Number of defects	3	0
Scrap	1	0
Bottleneck machine	Yes	
Availability	56%	80%
Changeover time	90 mins	
OEE	40%	85%
Number of operators	4	

Figure 3.6 A sample data box for process 1.

2. *What is the average quantity delivered?* This should help you to understand the labor required to unload the deliveries.
3. *What is the rate of customer orders?*
   a. *How many units must be manufactured per hour, per day, or per week to meet the customer demand (Takt rate)?* The operators will have an average production rate. If this is not known, it can be calculated or measured.
   b. *If we know the number of hours available for production and the number of units we must make in that time, we can calculate the Takt time, that is, how long it should take to make one unit.* This also will help us evaluate how many operators we need to achieve this rate.
   c. *How long does it take to use the incoming components (stock turns)?* This is used to help the purchasing department avoid buying too much at one time.
   d. *When we look for waste, we can work out how many units must be made per unit time and how many are actually made.*
4. *What is the rate of production?* How long does each process step take?
   a. *How many sellable-quality products are actually made per unit time?*
   b. *What is the capacity of the kit?* How many units should you be able to make in a given period of time if everything is running as specified? This will be available in the manufacturer's manual.

c. *By taking the ratio a:b, we can calculate the simple overall equipment efficiency (OEE).* The rate of production lets us calculate the *cycle time* for various stages of the process.
5. *Does the supplier have a minimum order?* It might be cheaper in the long term to buy fewer and pay a premium rate. This is not always an easy proof to make. There will be a lot of opposition based on perceived "need." The sales team (almost always) will want as much as possible in stores; purchasing can buy at a cheaper rate if it buys more; transportation may be free if a whole container is ordered.

   The downside is that stores will be overflowing onto the factory floor, warehouse personnel will be overworked, and the accountant will have cash-flow issues or a slower rate of return.
6. *How long does it take to get raw materials from order to delivery?* This has a major impact on your lead times and often defines your raw materials stock levels.
7. *How long does it take to unpack, check, and store the deliveries?* Measure and reduce.
8. *How many people does it take to unload and store the delivery?* For costing, as in point 7.
9. *When do the deliveries arrive?* Early morning, lunch time, at the end of the day, Friday night?
10. *What is the expected/specified quality?* Do you specify the quality of the goods? You want to know how reliable your suppliers are. Is the order complete?
11. *How do raw materials get to the operators to make the product?*
12. *How many shifts are there in the factory?*
13. *What equipment is used?*
    a. *How reliable is each piece of equipment?*
    b. *What is the availability?*
    c. *How much downtime is there?* How much is scheduled maintenance, and how much is due to breakdowns?
    d. *How long are machines down for changeovers?* How long for each, and how many?
    e. *Yield.* This is the ratio of the number of sellable units to the number actually made per unit time.
14. *How many production lines are there?*

15. *What is the capacity of each line for each product?* Capacity is linked to OEE. A line may have its capacity quoted based on how many it currently makes. The capacity is not what it makes but what it should be able to make in a given time period.
16. *How many tools are there for each process step?*
17. *How many people work at each process step?*
18. *Where are the bottlenecks?* See availability. The bottleneck also defines the speed of the total process.
19. *What is the batch size?* In Lean, we like to run smaller batches more frequently. This means that we need to improve changeover times, normally using SMED.
20. *How efficient is each process step?* The OEE.
21. *How long does each process take?* The process step lead time.
22. *Are the goods built and stored in a warehouse (built to stock) or as ordered by the customer (which is the preferred Lean goal)?*
23. *How much obsolete stock do you have?* This helps you to appreciate overproduction.
24. *What happens to obsolete stock?* How much of a loss is it? Despite having no chance of being sold, sometimes it is retained because it can increase the asset value of the company. How much does it cost to store it? Does it cause any other issues?
25. *How is product stored before shipping to customers?*
26. *How are the goods shipped to customers?* Late deliveries cost more.
27. *How long does it take to pack and ship the product?*
28. *Do you monitor supplier quality and reliability?* This has been mentioned a few times before. It is important to be able to depend on the incoming quality of the materials. Ideally, it should not be necessary for you to check them—that should be done by the supplier.
29. *Do you monitor your own quality?* How many faulty units do you make?
    a. *How many units do you have to scrap?*
    b. *How many units do you have to rework to get a sellable item?* How long does it take, and how much is that in cost?
    c. *How many orders are delivered to the customer late or with missing or wrong parts?* See on time and in full (OTIF).
30. *How does the customer place an order?*
31. *Do you have information issues when the orders are processed?* Are part numbers incorrect? Does the warehouse get the wrong parts? Are the

locations in the warehouse correct? Can the warehouse personnel find the parts?
32. *How is the production predicted/forecast?*
33. *How are the materials ordered?*
34. *How does goods-in know when the goods are to be delivered?* Do the goods-in personnel look out for orders not arriving or confirm that they will be arriving on time?
35. *Do parts get lost after delivery?* Some companies have flexible booking-in policies: Anyone takes it! I have experience with a few companies where emergency parts were delivered and accepted by the person nearest the door. The arrival is only discovered—days later—when the engineer chases up the urgent part to find out why it has not arrived.
36. *How are parts booked in?*
37. *How are they stored?*
38. *How do the production stages request material?*
39. *How does production prioritize or schedule jobs?*
40. *How does shipping know when goods are ready?*
41. *How does shipping arrange transport?*

On Time and In Full Deliveries (OTIF)

This is a very important parameter to track. Customer orders are not correct unless they arrive on time, on the original date, and with all the items ordered. Some companies claim to have a 100 percent delivery rate but count a rescheduled date as the original, provided that the customer has agreed to the delay. This is not a 100 percent OTIF.

Why is this so important? You are monitoring for problems with the production process. A late order tells you that something is wrong and has to be resolved. In addition, an unhappy customer will take his business elsewhere. His own production may be precisely planned and depend on the delivery, like the cream in the preceding example. Nonarrival could create losses for the customer.

From a cost perspective, late deliveries in Lean tend to be created by shortcomings in the processes—by the "7 wastes." The Scottish Manufacturing Advisory Service (SMAS) estimates the cost of late deliveries to be 1 percent of turnover for every 1 percent of late deliveries. It is assumed to level off at a success rate of 95 percent on time. I am not sure why. I have seen companies with delivery rates averaging only 50 percent, and the losses seem greater. Even so, 5 percent is not an amount to be scoffed at. Using a turnover of $1 million as an illustration, an OTIF of 95 percent would equate to an estimated loss of 5 percent, or $50,000.

Finding out which wastes cause the losses and eliminating them is an essential task. Whatever the actual value, if the OTIF is tracked, any sudden decrease suggests that a new problem has arisen somewhere in the process. The sooner the problem is discovered, the sooner it can be fixed, and the less the loss will be.

All the preceding information need not be entered into the data box. Most likely there will be around 10 key points. The goal is an understanding of the performance of the company, so you want to get numbers that relate to performance. Numbers can be compared and used to quantify losses and prioritize improvements. At a minimum, and because you are reviewing from a Lean perspective, you need to know how long a process step takes (the *cycle time*), the batch size made at each run, the best performance possible (how many units can be made in an hour or a day), the actual production rate, lost production time (the time to change from one process to another, the changeover time) and how reliable the equipment is (uptime and availability), available production hours (how many shifts in operation), the number of operators on each shift, and the main issues that reduce productivity.

Lead Times—Step 4

In order to plan production and deliveries, you need to know how long jobs take. The *lead time* is the time it takes to process a part from the end of one step to the end of the following step. Record the time in the dedicated box—a hexagon is the normal shape.

Many companies will quote a process time of, say, 60 minutes. They only count the time from the start to the end of the process and often automatically deduct any interruptions. By taking these two fixed points, you include any regular losses. The process step may take 60 minutes, but if it waits for two hours before the job starts, then the lead time is two hours *plus* 60 minutes. You want to make the factory process more efficient—to reduce the impact of the "7 wastes." You also want to reduce the two hours.

The lead time is a key target for improvement. Since you know that every process has two components, the part that *adds value* and the part that doesn't, the longer-term goal is to identify the wastes. This is the job of the process map. You also can consider streamlining the value in the process. There could be better ways to manufacture. By reducing the waste and then the process, you will make significant savings.

Inventory and Stocking Points—Step 5

Inventory includes *everything* in the factory that is used to make the product, including packaging and finished goods. An inventory point is identified using an *I* in a triangle.

You need to know the value of the raw material held in stores, how much finished stock is held, and how much stock is sitting on the production line as work-in-progress (WIP). This information can be entered as a number of items or as the time it takes to use that amount of material or both. Take care with the units, that is, minutes, hours, or days.

Questions for guidance:

1. How much raw material is held in stores?
2. How much raw material and partly assembled material is located at each workstation?
3. How much scrap is there?
4. Is there any obsolete stock?
5. How much material is waiting to be reworked?
6. How much unsold product is sitting in the warehouse as finished goods?
7. How long has it been there, and when do you expect to have sold it?

Quality Checkpoints—Step 6

Quality is crucial in Lean. Yet it is the odd-man-out. It is expected of everything you buy and is not optional. Quality checks do not add any extra value: The product already should be correct. It is "essential" non-value-added. Represent quality checks by a Q on the map.

Please don't get the wrong idea here. I believe that quality checks are essential. The point I wish to make is that excessive checking and poor quality cost money and personnel-hours. Bad quality costs customers. The customer expects quality—it is rarely an option. "I'll take six of those: four that work and two of the faulty ones." Even then, if the customer requests two faulty ones, she would expect them not to work. There are routine quality checks to ensure that the product does what it should, equipment calibrations, material standards, checks that are required by legislation, and standards that must be achieved to gain certifications. Certifications can add value.

So how can quality be a waste? When a machine is not performing properly, rather than fix it, or if it is too hard to fix, a quality check frequently is added to the process to tell whether another adjustment is needed. These checks remain a part of the process long after a permanent repair has been made. In Chapter 1, we discussed a manufacturer where the operators deliberately installed only three of the ten screws that hold the unit in place, faulty units were expected from the supplier, and steps were developed to accommodate them. Where a process is unreliable and produces a *variable* quality of output, don't accept the low yield—fix the process. It is not unusual to see (sleeping) operators sitting at bends in a conveyor where jams constantly stop flow and jars or bottles break. Would it not be more practical to change the track configuration to stop the jams?

Do you have too few or too many quality checks? Are they in the right places? There is no point in fitting hardware to a window frame if the next step scraps it owing to scratches on the glass from an earlier process. Does everyone know what they are actually looking for—what good quality looks like? Do you use go/no-go gauges to confirm sizes? For example, the part must pass through a hole to be correct as opposed to making time-consuming hand measurements? Are the problems of your own making, or are you buying them in?

End-of-line checks should not be the only check. At that point, all the damage is done, and you are left with only reworks or scrap, which costs a company more than twice the normal amount to make. One production manager told me a story about a precision-machined part. The first stage was to turn the diameter down to a very accurate size. During the next six weeks, holes were drilled, threads cuts, parts welded to the sides, the part even was placed in a furnace on two occasions to anneal the material, and finally, a second component was added. It was at this point that the company found out that the initial operation made the diameter too small. It was only out by one-thousandth of one inch, but that was too much. Six weeks were wasted, and all the material was scrapped.

In the case of underperforming equipment, fixing all the mechanical issues is not an easy option. Quality maintenance seems to be elusive. Even companies that believe in maintenance often do not provide the level required. The drive to make improvements will come only from the realization of how much the issues are costing in lost product, reworked materials, extra time taken, and additional quality checks.

Lean promotes the idea that operators do a job *right the first time*. If a product is made to the correct instructions, using the correct parts, and assembling them in the correct order—every time—the finished product will be made to the same quality every time. Each operator should confirm that her work is correct as it is being made *and* before it is passed to the next process step. Faulty units never should be passed on. The receiving operator should confirm that the work is correct before starting his work, which he also checks before passing it on. The number of defective parts should be tracked and graphed by the people doing the job. This will show where problems arise. Any problems should be prioritized for resolution. Lean also recommends that the operators are involved in developing the solutions. The operators' graphs of defect numbers will show when issues appear and if the fixes are working. The point of the graphs is short-interval control (SIC). The sooner a problem is identified, the less it costs the company.

For our map, we want to know where the safeguards are *now*. When we do a *future state map*, we can plan where we need them.

What Do We Need to Know?

1. *Is the quality of the goods ordered clearly specified?* In one engineering company, I watched a sales engineer draw a diagram of the customer's order from a phone call.
2. *Do operators know what good quality looks like and where mistakes are usually made?* Quality standards should be as absolute as possible, not left to the interpretation of the assessor. Use absolute tolerances, for example, ±1 millimeter, or have a photographic standard.
3. *Are incoming goods checked on delivery?* Don't wait until the part fails in use before finding out. Testing can be carried out on every unit or on random samples. Often the type of testing depends on the reliability of the supplier. But beware: Even the best companies can have problems owing to illness or whatever.
4. *What is the supplier failure rate?* Graph and monitor.
5. *Where else is the quality checked?*

6. Who checks the quality at each stage?
7. Is the testing equipment suitable?
8. Is the testing equipment calibrated?
9. How many quality failures do you have at each step and in total?
10. What is the customer complaint rate?
11. What are the main complaints?
12. What is your process for product improvement?

Quality is affected by job complexity, the standard of the instructions, the training level of the operators, and the quality of materials, tools, and equipment. Like everything else, we need to know what is wrong before we can fix it. Place a Q on the map to show the quality checkpoints.

Rework and Scrap Loops—Step 7

By default, quality issues and reworks are problems, so they should be added later. But at this point you want to get a feel for what the numbers are. Once you know, you can decide whether it is too much and what you can do to stop it. Try to identify numbers and a cash equivalent. The cash equivalent should include personnel-hours wasted to the point where it is scrapped *and* any personnel-hours wasted by people waiting downstream for the parts to do their job. If an hour of production time is lost, how many units could have been made in that time? You can work this out when you get to issues.

1. How much rework is there?
2. How much scrap is there?
3. Where are they generated?

What Are the Main Issues at Each Stage?

The map is almost complete. You have recorded all the "normal" process stages using yellow Post-its. Problems now should be added on *red* Post-its—*one* per problem. Start at the beginning of the map, and ask what goes wrong at each point. What stops people from doing what they should be doing?

I have not included a comprehensive list of prompts to ask the team. Many of them were covered earlier. Use the points that refer to your operation. If the data box lists a point, question what can go wrong. If there are 10 supplier complaints in a month, you need to ask, What are the complaints? How many of each type are there? How can we stop them? You want to identify the biggest issues. Every company has variations of the same problems whether they make pencils or heart pacemakers. Review the whole map and find the main problems using brainstorming—or as the politically correct would say, "Find the opportunities" at each stage.

1. What delivery issues do you have?
2. Is the warehouse cluttered and full?
3. Are any of the process steps unreliable? An unreliable process is identified by the variability of its output: in numbers and quality.
4. Is the process operation efficient?
5. Are the operators trained properly?
6. Does the equipment break down much?
7. What is your rate for OTIF deliveries? This should be monitored continuously and deviations investigated.
8. Is your production planning effective, or do you deliver to the customer who shouts the loudest?

Review every stage, and ask whether any of the "7 wastes" apply. In short, question everything. Even if you think that a step is correct, take a bit of time to discuss it. Even if you think that there is no other way to do something, question it. I regularly hear, "It is a biological, hand-crafted, engineered, textured, complicated, expert, black art, or whatever process, so the output *must* be that way." Well, in the odd case, it might be, but usually I find that it is not. This is one of the reasons that I recommend cross-functional teams. A person with no knowledge of a process will ask why something is done in a particular way when another with experience will *know* that it "has to be that way."

How Much Are the Problems Costing?

Identify the main issues, and find a way to quantify them. You can't fix everything, so you need to solve the top problems first. Lean refers to them as the "vital few." If problems are converted into numbers, it helps to prioritize the solutions. As a rough guide, fix the issues that cost the company the most money. The improvements will help to fund the process. To work out the cost, consider:

- ▲ How many components are scrapped?
- ▲ What is the scrap cost? A unit of product must have a value at each point of the process, so what is it?
- ▲ Does it have to be remade? If so, how long does it take, and how much is required in personnel, materials, machine time, and facilities?
- ▲ How long do employees have to wait until the production flow continues normally? How many of them? Convert the personnel-hours to cost.
- ▲ How much production is lost owing to waiting or rework?
- ▲ Will anyone need to work overtime at an enhanced rate?
- ▲ Will there be a special delivery or an air-freight charge or a contract penalty?

You might not get the cost exactly right, but whatever you get, it will be better than having no estimate. The value should be calculated as an annual amount. This will standardize costs and help you to appreciate the total impact of repeating issues. An issue may seem to have a low cost, say, only $100, but if it happens three times a week, 50 weeks a year, it really costs $15,000. The list will help you to decide which issues cost most in losses, how much they will cost to put right, and so what should be fixed first.

The Future State Map

The map that you just completed represents the way the process runs now. It is called a *current state map*. There is also a *future state map*. This is the map that represents how the process will look when the improvements have been made (Figure 3.7). Achieving this state is the strategic plan.

To create the future state map, you imagine what you want *and* what you know you can achieve, and then you redraw the current state map. In Figure 3.7, I took the liberty of assuming that process 2a was virtually all rework. Thus, by getting process 1 correct and balancing the other content between processes 1 and 3, process 2a could be eliminated. Other changes were made to quality inspections and inventory levels, waiting times were reduced, and process improvement targets were set.

A future state map is not always needed. However, it does emphasize the benefits and can include a *value-add plot*. However, I know and believe in the system. If I were a manager with a resistant team, I might choose to

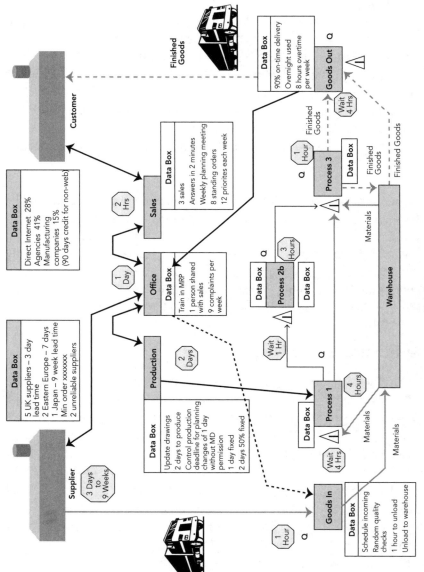

Figure 3-7 The future state map.

reinforce the project's success by creating the future state map. Worst case: It gives the team a bit more experience working with maps, which is not such a bad thing.

Solutions

You do not have to seek solutions here. It probably would be more appropriate to wait for the process map, where the issues can be understood in more detail. However, some obvious problems might be identified that will enable immediate savings. Highlight these fixes using blue Post-its. As always, there should be only *one* solution on each Post-it.

Much of the preceding methodology applies equally to the process map, as does finding the solutions. Look at every problem in turn. Use the "5 whys" and brainstorming to find a few possible solutions. If there is more than one possible solution, write the option on a second blue Post-it. If the problem is more complex, try a cause and effect diagram using added cards (CEDAC) or a problem-solving technique of your own choice (see Chapter 7).

Once you have a few possible solutions, try to put a number to them—a dollar value. How much will it cost to fix? How long will it take to fix, and how many people will it take? How hard will it be to fix this way? You are looking for a comparison between the problem cost and the cost of the fix.

How do you decide which problem to fix first? It is not only the amount of money it loses. I would suggest problems that repeat should be a priority, as should those which will cost very little to fix and can be done quickly. The ideal solution will have a high impact with a low cost to implement. Ignore issues that will cost a lot to fix and give very little improvement—unless there is some kind of safety issue or legislative requirement.

CHAPTER 4

Process Mapping

In Chapter 3 we looked at the big picture map. It has much in common with the process map. They both tend to ask the same questions because they are both looking for the same sorts of issues. The key difference in practice is in the detail being sought from the map. When diagnosing issues within an organization, the big picture map is the place to start. It helps you to decide where to concentrate resources and what to look for. In defining the types of questions to ask, I went into more detail than needed to save us from doing it twice. Rather than repeat all the instructions and Lean explanations here, please read Chapter 3 before this one.

Creating a *process map* is relatively simple. The layout is not rigid. You don't need to use the trucks and factory icons, but I do like to use the ones that work at the process level, such as a quality inspection, rework loop, or an inventory point. I recommend using the same basic color coding for all maps. In this way, as soon as you see any map anywhere in the company, you will immediately know what it means. I use yellow Post-it Notes for process steps, red for problems, and blue for ideas. I also attach copies of important documentation and add a small icon on each Post-it to show the step to which it refers (like a Windows text page as used in the copy icon, but an asterisk will do). I recommend that you do not draw any arrows linking Post-its until the map is complete. There is a high probability that the positions of the Post-its will be changed more than once as the map develops. It is easier to draw arrows or other symbols on Post-its that can be used temporarily and moved.

Both maps are designed to find problems and solutions *logically*. This sounds as easy as it is. The difference between the maps is the analysis level. As an analogy, if the map were a photograph, the big picture map would be an aerial shot of a town center showing the streets between the buildings; the process map, on the other hand, would zoom in on one house, making the detail of the house much clearer. It would show the doors and windows, the paths from front to rear, and any connections between adjacent houses and the garage, garden shed, and greenhouse.

In a factory situation, the big picture map reviews departments and equipment at a holistic level. Large processes are divided into *functional* sections containing a few machines or an automated line, where each part of the process defines a submodule. If preferred, each process line can be treated as a single unit. A data box will summarize the issues in the section. The process map, on the other hand, wants to know more. You would look closely at the operation of the department at individual task level, not just its inputs and outputs. You investigate how the material moves between storage areas and the operators. How the operators themselves interact. And you consider the equipment used and what you actually do to the materials or, if it is an administrative department, how the documentation flows. You want to find out where the flow is delayed or stops and where problems are created for others to fix down the line.

Bottlenecks

Here is one simple example, straight out of the "7 wastes." As you walk along a production line, every time you see a buildup of material (inventory) in front of a machine, it is telling you that something is wrong. You need to discover what. Is machine 1 making too much, at a rate faster than machine 2 can handle? There is no point in producing 100 items every hour on machine 1 if machine 2 can handle only 50 in an hour.

Perhaps machine 2 can handle the previous production—when it works properly. Here you could have a maintenance or calibration issue. Is it unreliable and regularly breaking down? Could there be too many unnecessary changeovers on machine 2, or is each changeover taking too long? Is machine 2 making more of an earlier product than it needs to satisfy current orders? Is there an operator on machine 2, or has he been called to help unload a truck? Maybe machine 2 works fine but is a bottleneck and just can't cope. (See the capacity map.) The Holy Grail of Lean is to have a production/assembly line where one item passes down the line from station to station without having to wait or without any of its operators having to wait for product—like a leaf following the flow of a stream.

I would like to make a quick point on bottlenecks. If you walk a production line in the *reverse* direction, that is, from the end to the start, the first buildup of inventory is the bottleneck. This could be the one causing inventory buildups all the way back along the line.

Mapping a Restaurant Process

Let's get back to the maps. Suppose that you are running a restaurant and want to consider the initial customer interaction. For this, a big picture map "process" would consist of around three major steps, but a process map could have 12 or more steps. Table 4.1 illustrates a comparison.

Table 4.1 Restaurant Operation

Big Picture Map	Process Map
1. Check that the table is ready.	1. Ensure that the tablecloth is clean.
	2. Ensure that the plates are clean.
	3. Ensure that all the plates are set out.
	4. Ensure that the glasses are set out.
	5. Ensure that the cutlery is set out and is clean.
	6. Ensure that the bread has been placed.
	7. Check that salt, pepper, sugar, and sweeteners are topped up.
2. Take the order from the customers.	8. Show the customers to the table.
	9. Seat the customers.
	10. Hand the customers menus.
	11. Let the customers decide what they want to order.
3. Record the order in your notebook.	12. Record the order in your notebook.

In the table, the big picture map lists only three steps. It could just as easily have been only one—"Take the order from the customers" because it implies in one statement that the first two steps have taken place and the customers are at the table. So why bother with having 11 extra steps as we do in the process map?

In the process map, listing the details enables you to consider how everything is done and what can be done to offer the customers better service. It is not always the big issues that cause the greatest problems. You want to avoid all the issues. More important, you want to find and eliminate issues that repeat time and time again. Steps 1 through 7, as it turns out, are all quality inspections.

1. The customer will expect a clean tablecloth. If you regard a dirty cloth as a *defect*, you would want to know how often it happens. How often are the tablecloths cleaned? Do you ever have to wait for clean cloths? Do you have any storage issues? Could you introduce mats to cut down on cleaning but not sacrifice quality? Is there a better tablecloth you could use, perhaps one that can be wiped clean?
2. Do you ever get the place settings wrong? Is there a training issue? Are there enough plates?
3. Are the glasses in the correct places? Are they clean? Do we have a washing up problem?

-
-
-

12. Are the orders written or memorized? How often do you get the orders wrong?

I do not intend to go through all the steps. I just want to illustrate a point. If you look at the problem in detail and apply the "7 wastes" to each step, there very likely will be a number of things that you can improve. A faster table turnaround could mean an extra meal at each table in an evening, which would mean increased turnover.

The Team

The big picture map is more strategic and is best suited to managers. When planning a process map, you want different information, so how do you decide who should be on the team? Like most improvement facilitators, I am an advocate of cross-functional teams. For the first map you carry out, it would be wise to use people who you know will perform well. In the past, I would include an "awkward cuss" on all my teams. When I won that person over, I had a convert to the process that no one would have expected. This was real proof that the process really works! Now I lean toward a willing team. Having limited time, implementation becomes a bigger issue. You cannot afford the luxury of the extra "conversion" time. The process will progress more smoothly and even may have a better outcome.

Note that I said *may* have a better outcome. I believe that the awkward team members have a lot to contribute. They act as catalysts for the others and get them started because they are not afraid to ask questions. When looking for problems, the team needs to feel confident enough to point out issues that do not work. A team of "yes men" might only stroke egos, agree with the boss, and avoid the difficult issues. James Bond's vodka martini has to be shaken, not stirred, to get the best flavor. To get the best outcome from my teams, I like a bit more interaction. Even so, this has to be balanced with the situation if you are running the process on your own.

Perhaps I think this way because I can be a bit awkward, too. Coming from a science background, I was taught to question everything. I was once promoted because I questioned an experiment a senior lecturer was developing. He left the room at speed and went to talk to the department professor. When he returned, he told me that he was not going to be corrected by someone of my grade—and told me that I was being promoted.

Perhaps you don't need the immediate or continual hassle. The dissenters always can be brought in at the end of the analysis, as a group, and asked to give their input. The interesting point is that their input will be worthwhile. Thus who should be included on the team? I like to average six people. I don't agree with having large numbers. There can be too much discussion and too few *agreed-on* solutions. I also disagree with the theory that the best team has three members, two of whom can't make it to the meetings! Six is a good number if it is not too much of a drain on production; if it is, three will do.

The key member is the *team leader/facilitator*. She should know the mapping process and can guide the team. It would help if she also were able to train the others in the process. A couple of operators from the production line under analysis will know how the line really operates. They will know the difference between the written process and what is *actually* accepted. An operator from another line can add his experience. An engineer or maintenance technician is useful as someone who understands the operation of the equipment. She will have input on equipment performance, availability, and quality—the building blocks of overall equipment efficiency (OEE). I once had a forklift driver on a team. He was nothing short of excellent.

Remarkably, no one had asked his opinion before. (He has since been promoted.) Consider someone from the office, the design department, or purchasing. They also will have an input. It is very, very useful to have access to someone from finance. Such a person is invaluable in converting issues into cash equivalents, a very useful skill and a primary goal of the analysis.

Some teams are unhappy discussing problems as a cash equivalent. I don't do it just for the money. It is a common denominator for standardization: Scrap and rework have a cost in labor, as do replacement parts; waiting for operators has a cost in lost production and wasted labor; broken equipment costs in almost every area. Teams don't feel so badly, however, when at the end of a process they have saved the company oodles of money!

Where staff numbers are limited and forming a team could have a negative impact on production, it is possible to opt for a core team with extra support. Two or three permanent members can maintain continuity, and others can be called into meetings as required or on a rotation. I have found that this works quite well. It also helps to keep meeting times flexible, but ensure that there is a window within which meetings must be held. Otherwise, the meetings will stop as soon as production has problems. It even could be worth keeping an attendance register. Even if the managing director has bought into the process, the production manager often still gets judged on the numbers at the end of the line! If a complex issue is discovered, it is worth considering the input of a vendor engineer. Some companies will provide free support because it can strengthen customer/supplier bonds, and any solution may mean future equipment sales.

The team must have introductory training in process mapping and Lean theory, particularly the "7 wastes."

Creating the Map

Creating the map is much the same as in Chapter 3, except that you don't have such a formal layout. Having said that, some people have used the process map format to create big picture maps. Lots of teams stick the Post-its to rolls of brown paper. Indeed, you will hear Lean practitioners frequently refer to "brown paper maps." However, it is equally possible to use white wallpaper. One sign maker I knew used a roll of high-quality printing paper. Sticking the Post-its onto the paper protects the wall and allows the entire map to be taken off the wall and stored. Others work directly on the wall or on glass partitions. Leaving the map on the wall invites everyone in the factory to see what is happening and prevents the rumor mill from starting up.

If you should create the map in a public meeting room, you might not want visitors to see any issues. To that end, a map that can be taken down is a benefit. Other companies set aside a "war room" solely for the use of improvement groups. My own preference is another paper format. I recently had a revelation. I discovered that Post-its also sell a sticky-backed flip chart pad. Now, rather than use a single, long sheet, I use these sheets like tiles, sticking them in a row to the length I need. The "tiles" make introducing a missing section of a process so much easier. Rather than moving every individual Post-it to the right, they allow complete "pages" to be moved. They also allows sheets to be added above or below the main process line, accommodating parallel processes or options.

Trick number 1: As shown in Figure 4.1, don't start the map at the left edge of the wall. Place the first process step several feet away from the edge. It will not take long before the process issues spill backwards and fill the space.

I cannot discuss specific customers' maps, so I have represented the Post-its as rectangles with a letter to show what color it is (R = red; Y = yellow; B = blue). The figure represents a main process having a couple of options. It is the same process type as the "arrow" map represented in Figure 4.2d. Ignoring the administrative space on the left side, you can see options following steps 1, 8, and 14. You can see that there is a choice of two options after step 14. This process could be a simple mobile phone with the option of different internal memory sizes, screen sizes, case color, and supplied extras, such as earphones or a car charger.

To assemble the map, assume that the first step chosen is: "Set up the job." This is not enough detail, so what if we change it to a more specific step: "Prepare the equipment." Even this is too general. As a part of this operation, the operator may need to perform a changeover. But what do we need to change to? This reduces the options for first step to: "What is the next job?"

I can't tell you how often I go to a company where the operators have no idea what their next job will be. A primary goal of the production manager should be to have the current day's jobs for all equipment and production lines planned in advance. This schedule should be fixed, with

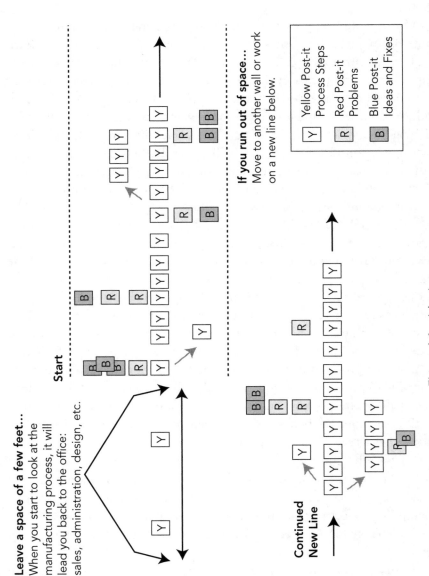

Figure 4.1 A basic process map.

changes being rare and only for absolute emergencies. Better still, the jobs should be on a production list or board located at each workstation. Having advance knowledge of the next job means that it can be organized in advance. The first eight process steps now might become:

1. Find out the next job.
2. Collect the changeover parts.
3. Remove the existing parts.
4. Install the new parts.
5. Collect the materials from stores.
6. Prepare to start the run.
7. Confirm that everything is ready to run.
8. Start the run.

I have listed only the first eight steps, but later, when the entire process is complete, it will be revisited by the team, starting at step 1, and be analyzed for problems.

Let's consider the potential problems for step 1: "Find out the next job." The most common complaint is that the operator rarely knows what the next job is, and it takes her ages to find out. She frequently complains that the job sheets are not written or are incomplete. Another complaint is when she has to stand around and wait for production to decide what to make next. These are not the only issues. It is best if you can find out as many of the reasons as you can. You then can look for common causes and which issues you can improve. Already, from the three issues listed, you have a suspicion that the planning office is not operating efficiently.

While we are discussing finding the next job, it is worth mentioning a practice that causes late deliveries. In this example, all the current day's jobs are left in a pile, and the operator can select the one he wants to do. It is human nature for him to select an easy job, leaving the more difficult ones to the end or to the following day. Working this way makes it impossible to plan any setup in advance; no one knows what the next job is until the operator picks one. The schedule should be organized like a school timetable, each subject completed in sequence, with the chosen teacher, planned to have an exact start and end time. It is only in this way that you can spot delays and the issues that cause them. If you don't see problems, you can't fix them.

One of the preferred equipment performance figures to track is the *cycle time*. This is deliberately defined as the time taken from the *end* of one process to the *end* of the next process. Everything that increases the cycle time unnecessarily is a waste. You know intuitively that any delays in performing the tasks are wastes. These include both having to wait for instructions and the delays in tracking down information. Assuming that the company has no problems selling its product, all the time wasted has a corresponding productivity loss and a cash equivalent.

To find as many issues as possible, brainstorm each step using the "7 wastes" as prompts, and look for the broader issues. The *W* in TIM WOOD relates to waiting.

"What do you wait for?" One answer to this question is information. The fact that the operator has to stop her work to find out what to do next is a problem in its own right. This can be resolved only by production planning. One acceptable solution is a board at the equipment with the next three jobs listed in order. In this way, the operator always knows what job is next, and stores can be coordinated to ensure that the parts are ready. Initially, to start the improvement process, you would settle for one job in advance!

To avoid repetition in writing this section, I will consider all eight steps together because it is an imaginary example and my objective is to teach the process. I know what I want you to get out of it. Besides, most of you will not work to the individual step level and will take shortcuts. This is not a big problem because missing issues probably will show up again when another process is analyzed.

Consider the Impact of the "7 Wastes" on the First Eight Steps as Defined Earlier

Transporting means having to carry goods around or search remote offices for needed information. There are issues in process steps 1, 2, and 5. The operator never should have to leave the tool to get information. A proper planning process will ensure that he has everything he needs to do the job *before* it is needed.

The changeover parts in step 2 could be a transporting waste if there is a large distance to the storage area and if a number of items need to be found. Those items could be stored closer to the machine. You want easy access, like taking a book from a shelf. More often than not, parts should be stored as close as is practical to where they are needed or be delivered to the operator just *before* she needs them.

In step 5, again depending on where the stores are located, you could have a transporting waste. Lean frequently positions the material "supermarkets" close to equipment, where a small quantity of material can be left for short-term use. This store can be in the form of shelves or, better still, a gravity system that feeds forward as the material is used. Once more, the goal is for the materials to be delivered to the operator just *before* he needs them. Where a "supermarket" is used, this should be topped up by the warehouse person or the material handler *before* it runs out and be changed to the next job's materials as required just *before* the operator needs it. In this way, the operator does not need to leave the tool.

Did you notice that every "before" was highlighted? Waiting is one of the biggest offenders. The "just before" is to avoid delivering the items too soon and creating any inventory issues or physical obstructions.

Inventory covers everything from raw materials to work-in-progress (WIP) to finished goods. It creates issues in step 5 and possibly step 2 if the changeover parts were stored in a faulty condition and need repair parts. When parts are delivered to the operator, plan a balance between preventing the operator running out of material, having the warehouse person make too many trips from the warehouse, and keeping too much material sitting next to the machine. You do not want to block passages or take up valuable floor space that could be used for a changeover rack or by another process.

In Chapter 3 I recommended a two-box *Kanban* system to prevent stored parts from running out. This system also can be used to control inventory levels at equipment and workstations. Depending on the size of the components, the operator is given two lots of material for use. It can be delivered in boxes or on pallets. When the first box or pallet has been used, the operator switches to the second. The empty pallet or box is the signal for more material to be delivered. The two-box *Kanban* avoids multiple boxes being dropped off at a tool. Large amounts of material can be used, but remember, any time the warehouse person or forklift driver is spending delivering parts to a machine that will not used for hours, he is not doing something else that is needed now.

Movement refers to the workstations. The goal is to design a layout such that as much of the equipment and materials needed by the user is within easy reach. When explaining this feature, I often use the drummer or keyboard player of a rock band as an example. In both cases, it is possible to play every instrument or drum from the central position. In the case of the keyboard player, she can play multiple instruments at once.

There might be movement issues in step 2, 3, or 4 if the changeover parts are hard to reach either in stores or during the changeover. If fitting the parts is difficult, there could be a saving by using two operators to do the job. The first would make the change, and the second would feed the parts. The goal is not always to do less work but to organize the workload in such a way as to get the equipment back into production as quickly as possible.

In the case of a changeover, using the single-minute exchange of die (SMED) process will reduce the time it takes to complete the task and also will help to redesign the way the changeover is done.

Waiting is one of the biggest wastes. It affects every step. Not having instructions on the next job or having bad instructions delays the entire process. This is worse if the production department is waiting to find out what to do. Arguably, the only "acceptable" situation is in the case of a very demanding but important customer who keeps changing her mind and ordering at short notice. In the past, where this has happened to my clients, I recommend planning a slot for them as a company but not specifying what is to be made. If no order is placed, you can run the next job on the list. If the customer calls with her "new" emergency, you simply fit it into the slot. It is possible to regain some control: Proactively call the customer daily to check that she has no urgent requests. This might provide a few hours of extra notice.

Collecting the changeover parts is also waiting because the next job cannot start until the change is complete. These parts can be delivered in advance by the material handler while the operator is running production. This also holds true for collecting the raw materials from stores.

Less acceptable and completely avoidable is when you have to wait for the changeover parts to be repaired. The parts never should be stored in a damaged condition, which is, sadly, not uncommon. Equipment reliability also can cause waiting. The issues will surface as the machine is being set up for the next job or when the quality check shows that more adjustment is needed for alignment or speed. The solution is to ensure that all parts are correct before being stored and to carry out preventive maintenance that ensures that the alignment and setup controls are accurate.

Overprocessing refers to carrying out process steps that are no longer necessary. This is the waste I find most difficult because it involves questioning the operation of the process. In step 5, where you confirm that everything is ready to run, if the raw materials comply with their specifications, if the machine is running properly, and if the control dials are accurate and the setup procedure is precise, there should be no reason for the process not to run first time. If it fails to run, you need to analyze what failed and find an idiot-proof modification to prevent it in future. I mentioned the process where the incoming product was counted and then scanned. The counting is overprocessing because the scanning also counts. Other examples would be providing a 4-millimeter metal panel when 2 or 3 millimeters would suffice, spray painting extra coats of paint that will not improve the operation or appearance of the unit, and printing a billboard to be sited halfway up a building at 1,200 dots per inch when 600 dots per inch is acceptable for printing photographs. (*Note:* A billboard is rarely viewed at arm's length.)

One customer with whom I worked noticed that the first thing his production operators did at the start of a shift was reset the dials to their favorite positions. This could spoil a perfectly acceptable run. To resolve the situation, he modified the control panel of the machine by disconnecting the controls that created the worst damage. These changes now had to be made from inside the unit. This was a drastic solution, but it was very effective!

Overproduction is making too many of an item. This is an interesting waste because different departments all will justify why their need to have more or make more outweighs the other issues. Sales wants to have the raw materials so that it can promise immediate delivery or never refuse an order. The same argument is used for finished goods. Purchasing likes it because it can get bulk discounts. Production wants to make as much as possible while the machine is set up because it saves changeover time and compensates for the fact that the machine fails a lot. Many operators find that running the machine is easier than carrying out changeovers. Maintenance wants to make as much as possible, too, because it takes the pressure off its workers to improve equipment performance. The managing director often acquiesces because it makes her life easier. She will not be too happy, though, once she has a better appreciation of the increased costs. The only people who absolutely do not want all this stuff are the warehouse people, who need to juggle everything and control space, and the accountant, who has to find the money to pay for it and all the issues it causes.

We need to discuss what defines *too many*. Again, it is a balancing act. The technique is called *production leveling*. Overproduction has no impact on the first eight process steps other than the possibility that the new job may require too many parts to be made. If there are too many, this will have an impact on the length of the production run, possibly causing subsequent products to be delivered late to the customer.

The fourth principle of Lean is *pull*—make what your customer wants to buy. In this way, you make an exact amount of product. Unremarkably, the opposite of pull is push. *Push* occurs when an amount of product is made before customers are found to buy it.

Most of us have to forecast what we can sell. Few of us consider that another word for forecast is *guess*. A few companies live in Lean Nirvana; they *make to order*, which means that they only ever manufacture what customers want. They tend to make bespoke products that are specific to people or their property—some precision engineering companies, double glazing, top-of-the-range doors, special birthday cakes, made-to-measure suits, (some) school uniforms, brochure and poster printers, and repair companies that provide builders, plumbers, joiners, and roofers. Yet even these companies need to estimate the materials they need to keep in stock to supply their demand. They can't all be like the local handyman, who "needs to nip down to the do-it-yourself shop" for the part he needs.

It is hard to overcome the temptation for companies to make as many units as they can "while the machine is set up." In the days of mass production, getting as much as possible out of a machine was a goal. It was called *maximizing utilization*. It did not matter if there were no customers, as long as the machine ran for a full day. Overproduction is a mixed blessing. If you have oodles of stock in store, bad manufacturing practices are hidden. It is not so important when *something* goes wrong. After all, a loss of a few hours of production has been allowed for by making extra.

The closer we get to efficient production, the more reliable the processes need to be. The downsides of overproduction outweigh the benefits. These tend to be a need for large warehouses, often including external storage space, that has to be paid for; a cash-flow problem, caused by buying more material than needed; labor wasted in making units just to sit on shelves; a high risk of old stock becoming obsolete; and individual production runs taking much longer than they should. Consider this: If your company has an overstocked warehouse and external storage, would doubling the size of the warehouse solve the problem? Or would it simply introduce some breathing space until the new warehouse is filled up?

A production manager told me a story about a previous employer. The company bought two container loads of material from the Far East at a very competitive price. When the material arrived, the company had no place to put it. As luck would have it, the company was able to hire unused factory space close by. The goods were unloaded into the factory but were not indexed, so it was difficult to find the required materials. In the end, the company returned to ordering new material as the factory needed it, just as it had done before, but it continued to pay the extra rent.

Consider another company that buys raw materials in large quantities—as a number of pallets. One customer's demand, after analyzing previous orders (ideally over a three-year period), is shown to be an average of five pallets in a month, with a maximum of 10 in a month. Yet, owing to discounting, the purchasing department buys 30 pallets of raw materials at a time. This is enough to supply the customer for half a year. In the situation I witnessed, the company chose to make all 30 pallets worth of goods. There was no savings in storage because pallets of raw materials simply became pallets of finished goods. The major issue was a key production machine being tied up for a full five days, preventing anything else from being made that was needed now.

How likely would it have been for the customer to phone up and ask for an extra 25 pallets worth of stock? Zero! What about the chance of a sudden order for an extra 10, 15, or 20? Management discussion suggested that 10 pallets could be possible, so maybe a reasonable number to have in stock would be 15 units—the average plus the biggest extra order. The company meeting didn't include the customer, the main person in the discussion and the one you want to make sure gets what she wants.

Just as the supplier has a maximum capacity that it can make, the customer will have a maximum usage rate. What if the customer says that she can use only five pallets in a week, no matter what her demand? Now we know that we need to carry only an extra five pallets to cover an emergency and have a full week to make any extra before the customer has an issue. You have just won back three full days of production time.

Extra stock sits in the warehouse fully paid for. A cost comparison must be carried out to consider all the consequences of buying so much against the amount of discount sacrificed. It even may be possible to negotiate a deal with the supplier to get the bulk price (or close to it) based on confirmed orders over a six-month period. The old joke is funny because it is true: My wife does not work for a purchasing department, but she still comes home with new shoes or a jacket she has bought, telling me how much she has saved by buying something she never intended to buy in the first place.

A *defect* is anything that is *not right the first time*. A defect will lead to scrap or to a rework being needed. Defects are one of the two biggest wastes. I talk about right the first time, but the former definition is more correct. It highlights everything that has to be put right or be scrapped. Which of the eight can suffer from defects? All of them!

In steps 1 to 8, "the next job" could have *defective* information or missing data. The changeover parts could be damaged. It is not unusual for faulty parts to be removed from equipment and put in stores without being repaired or even telling anyone they need repairs. The new parts can be installed incorrectly, the raw materials can be loaded in the wrong orientation, or the machine alignment may not be set up properly. The raw materials may not be in stores, can't be found, may be damaged, may be faulty, or the wrong parts may be delivered to the machine. Steps 6 through 8 relate to the time it takes to tweak the machine so that it actually runs production. This is why the changeover time of a machine is not only the time to carry out the physical change but also is measured from the *end* of the previous product run to the first *good* run of the new product. If the changeover procedure is well organized, minimal tweaking is required.

It is essential to brainstorm each of the steps to identify what the issues are and to find the recurring problems. If there is a production log, all issues should be recorded there for future analysis. There will be small issues that people do not feel are worth recording. Here, you can use a basic tick sheet or put a dot on a photo of the area where the minor stops occur. When it is time to analyze, you simply count the ticks or the dots. Remember, you are looking for symptoms that will lead to root-cause issues. The problems you discover often will lead back to sales, order taking, job preparation, design, instructions and drawings, and the planning process.

The process map can end up having any number of different shapes, as can be seen in Figure 4.2. You could have a process with three input stages, making parts that are all needed before the final unit can be completely assembled. If the product is, say, a washing machine, line 1 might be making the rotating-drum assembly and adding the water pipes and valves. Line 2

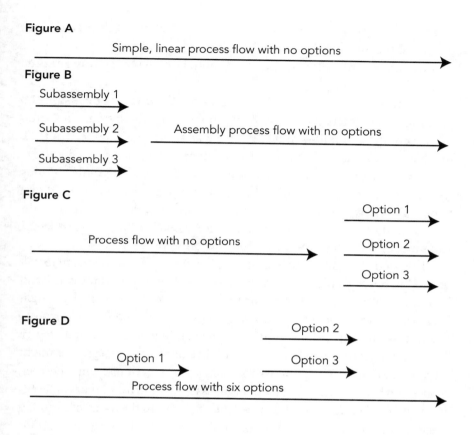

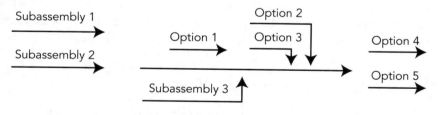

Subassemblies are designed to join the process where needed. The trigger point for starting a subassembly can be a step in the process, as in Figure E. Subassemblies as illustrated in these examples could also be replaced by options.

Figure 4.2 Types of process layouts.

might be preparing the wiring loom and assembling the electronic control panel and the motor. Line 3 might be making the chassis and outer panels. All three are needed before the final assembly, although they would appear at different points on the line.

As you progress through the chassis line, there might be a split for painting panels different colors or drilling extra holes for options, and some of the product will follow route A, some will divert to route B or route C, or whatever. In this case, you simply introduce a new line of Post-its above or below the main line. If possible, leave space for problems and solutions (red and blue Post-its). At this point, I must repeat a warning: Do not be tempted to draw any lines on the map—unless they can be erased easily. Wait until the map is complete and no missing entries have to be added. If you must include an arrow, then draw one on a Post-it.

The process detail is also a balancing act. Too detailed, and the map will be too big but will still work; not enough detail, and important issues can be missed. New teams tend toward steps that are not detailed enough. However, if thorough enough, questioning for issues (red Post-its) should flag important steps, allowing them to be subdivided. If the first step is: "Load the machine," you bypass the first five of the eight steps. You might miss some of the preparation issues—getting instructions from wherever, checking whether there are any obvious issues, confirming that materials are available, determining who brings the materials to the tool, determining whether the tool needs to be set up, confirming that there are instructions for the setup, doing the setup work right the first time, determining where the setup parts are located, and determining how long it takes to set up the tool and whether the tool needs to run a while to warm up. If the steps are too big, each yellow Post-it will have a whole lot of red ones above it.

In Summary: Making the Map

Follow the same philosophy you used in creating the big picture map.

1. Choose the process you wish to analyze.
2. List the steps in order using yellow Post-its.
3. When complete, return to step 1 and start looking for issues that delay or prevent product from being made. Each issue should be recorded on a red Post-it. As a prompt for questioning, use the "7 wastes" as a guide.

4. When all steps are complete, return to step 1 and, with the help of someone from finance, estimate the cost of each problem, as explained earlier. Standardize the estimate over a year to take account of the total cost of recurring issues. It is worth creating a simple spreadsheet to record the data and how the calculations have been made. Since many of the areas of loss will be the same, adding some simple equations will generate the answers for you.
5. Identify the biggest issues—the vital few.
6. Brainstorm for possible solutions. Each idea should be recorded on a blue Post-it.
7. Time permitting, evaluate the impact of all the problems. Use the spreadsheet created in step 4.
8. Consider resolving some of the smaller issues. Many of the smaller problems will be easy to fix. They only exist because no one bothers about them. Now that you know what they are costing the company, ignoring them may not be so simple.

The Capacity Map

The *capacity map* compares production rates, but the same details can help in a process map. When a machine or an operation is listed, it helps to list some productivity details as found in a data box. For example, "Cut the panel to size" might record how many panels can be made in an hour. Where a job is a "one off," where is the estimate of how long the job should take? Work on averages. You are looking for variations. When you expect a job to take 30 minutes and it takes 45 or 60 minutes, you must ask why. If you expect to make 100 units in a shift but only get 80 units, you must ask why.

Problems often happen because people make wrong assumptions. Managers, after discussing a new job with a customer, assume that an operator or an engineer should *know* what to do and that she only needs a brief set of instructions. "She should have known what to do; it is exactly the same as we did last year for Bash It & Co." The average person needs to repeat a job around 17 times before it becomes automatic. I have found that if I do not run a course for just a few months, I need to refresh the detail before doing another. Ensuring right information can seem time-consuming, but there is no alternative.

Remember, you are looking for *root-cause* issues to fix. Root causes for equipment can be applied, with a bit of imagination, to all processes. The Japan Institute of Plant Maintenance (JIPM) in its analysis for total productive maintenance (TPM) identified 16 causes for poor production performance. These can be reduced to three main areas: people, equipment, and materials. A machine that breaks down is a symptom. The root cause could be due to:

1. Poor or no maintenance
2. Poor or no instructions
3. Poor or no training
4. Poor quality of materials
5. Operating the equipment incorrectly

The 4M's are another source of prompts, often used as a base configuration when making fishbone diagrams. Basically, all problems can be boiled down to four causes: man, machine, materials, or methods. They are clearly echoed in steps 1 to 5 above. My wife would disagree with the 4M's. She reckons men are responsible for all problems.

Referring to the TPM prompts above, wrong or no parts for a process could be due to instructions, training, or materials or having no *method* for maintaining the store. An operator making a mistake in a dimension could be the result of maintenance, instructions, training, or equipment operations or man, machine, or method. Not getting all the details of an order could be due to instructions or training or man or method. You need to find and fix the root cause if you want a permanent solution.

CHAPTER 5

Capacity Mapping

My job with the Scottish Manufacturing Advisory Service (SMAS) is as a Lean practitioner. For all intents and purposes, I am an improvement consultant. But I do not just apply Lean methodologies. My technique has evolved from my education, my work experience, and techniques that I have learned from friends, colleagues, and books. One thing I have learned that I want to pass on to you is that virtually all the techniques follow very similar lines. It was this feature that led me to develop my system of blending processes to suit the needs of the client.

The Unloading Process

It came as a surprise, then, when I did a process map for a client, and it showed only two problems. The process was introduced in Chapter 2. It was probably the shortest process I have ever mapped. I cannot describe the actual process for reasons of confidentiality, but I can create a facsimile. The process steps are as follows and are illustrated in Figure 5.1:

1. Deliver the product to the factory.
2. Transfer the materials to the loading area.
3. Transfer the materials into the processing drum. The drum has an output to a conveyor belt, where the quality is checked as the material is transferred.
4. Move the materials by belt into a machine that squashes them into predefined shapes.
5. Sort the bales by content and shape.
6. Store the bales.
7. Sell the bales.

8. Load the bales into vehicles.
9. Ship the bales to the customer.

The two process problems originally seen were variable amounts of contamination from the suppliers of the incoming material and the shaper/baler needing a maintenance schedule. Neither was significant, and yet the process was not running anywhere near as efficiently as it should be. My colleague, Colin Allan, and I stared at the map, looking for inspiration. The team sat silently in anticipation of guidance.

The Birth of the Capacity Map

While looking at the map, my mind's eye overlaid an electronic diagram. It was a basic building block that became the state of the art in electronics in the 1970s, when microchips were being developed. It's called an *operational amplifier*. For those of you with no electronics knowledge, please don't be put off at this point. The operational amplifier revolutionized electronics. Early sound amplifiers were complicated circuits with *hundreds* of components. Microelectronics reduced all the components into a single chip about half the size of a thumbnail. The clever bit was that all the user had to do was add a couple of components to feed the electrical "sound" to the amplifier and a component to set the *gain* (how loud the amplifier makes the sound). In the case of an electric guitar amp, for example, all that was needed was a plug socket, a volume dial (unless it was a rock band, where maximum was the only setting used), and an output for wiring to the speakers.

Operational amplifiers are still available 40 years later but are much, much more sophisticated. They still do exactly what the name suggests—amplify the input and make it bigger or louder. We use them to amplify the signal from TV aerials, sound in TVs and sound systems, radios, phones, computers, MP3 players, guitars, and microphones in bands and CD players. In short, we use them for amplifying everything.

So what? Well, the thing about these amplifiers is that they can only amplify sound to the level of the battery that powers them. They all have a maximum output. In the case of sound, they have a maximum volume. This brings us back to the process. If you look at Figure 5.2, you can see the diagram of the amplifier sitting below the process map.

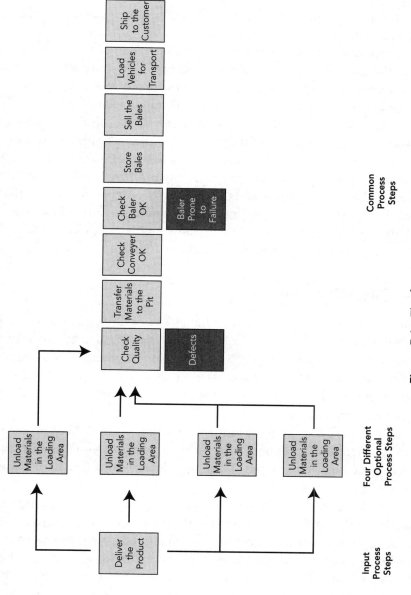

Figure 5.1 The four-input process.

The amplifier has a series of inputs in the same configuration as the process. The main process is not shown as eight steps but rather as one triangle—the shape being completely irrelevant from a practical point of view. The amplifier has a fixed output volume, or capacity, as does the baler, which has a maximum production rate.

In both cases, the output is *limited* by the operation of the input. In the case of the amplifier, an input dial with fixed gains ($R1$, $R2$, or $R3$, operating like a volume control) limits the output. In the case of the process, it is the way the material is delivered and unloaded that limits the output.

We have all heard about bottlenecks and how they limit production. We also have an appreciation of the theory of constraints, although some of us might not be aware of the technical name. It is best explained in Eli Goldratt's book, *The Goal*. The *capacity map* is a simple way to identify process and equipment constraints. It helps us to recognize operations that have limits *and* those which are limited by the processes around them. For analysis, we will use basic Lean theory and, sometimes, more technical processes such as overall equipment efficiency (OEE).

The Capacity Map

If this technique already exists, then I apologize. I do not intend to steal anyone's idea. As explained earlier, I *found* my capacity map by necessity. I was looking at the map in Figure 5.1, a really simple process map that had only two red Post-it Notes (problems). Yet I knew that the process was not efficient: It should have been much better. It was just not clear where the losses were coming from.

So, as I stared at the process map, I recognized that the four input options acted like different resistances—*constraints*—to the process achieving its maximum output. The larger the resistance, the fewer units got out of the process.

Consider a crowd of people passing through a door. The bigger the door, the smaller is the resistance to the flow of people passing through it. This allows more people to pass through in a given time. The size of the door limits its capacity. Another example is the size of a window. A big window lets more light through than a smaller window, so it follows the same argument as the door. Additionally, bad organization can make

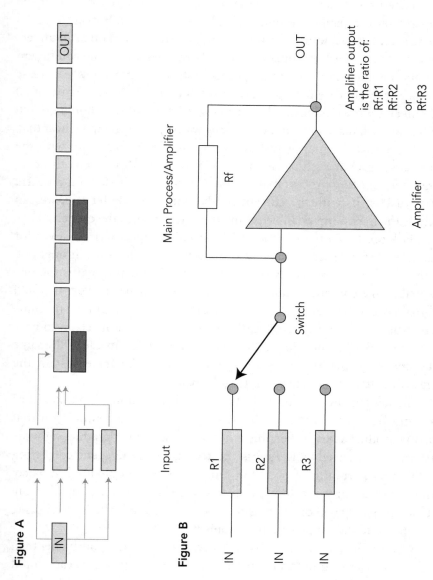

Figure 5.2 Comparing the process with the amplifier circuit.

matters worse. If you position something in front of the window, you add to the *resistance* and reduce the amount of light transmitted even more.

In the process under consideration, the operators reported that the shaper/baler is capable of producing 30 bales in an hour. Initially, I would always accept the value given by the operators, although it is often an underestimate. You can ask the operator later, if required, to get the equipment operation manual and check the specified rate, or you can slowly increase targets as the improvement process becomes established. The figure of 30 bales in an hour would mean that in an eight-hour day, you should get 240 bales from each machine. But the company wasn't getting anything near that.

Notice that the map is just an evolution of a process map with the production rate at each stage added (Figure 5.3). All we need to do to convert an existing process map into a capacity map is stick on an extra series of Post-its with the current production rates. Better still, we can include the maximum possible production rates, that is, the capacity.

Mark on the Post-it whether it is a specified capacity (from a vendor manual), a best rate achieved, or an estimate based on the usual throughput. In the example, there is only one *real* capacity limit, and that is the one defined by the equipment manufacturer. The shaper/baler's Post-it would be marked as a "best-ever rate." We will assume the capacity of the machine to be true because it was the best throughput the operators claim to have ever achieved. By association, other steps also must be capable of the same rate: checking quality, filling the drum, moving the materials on the conveyor belt, and the rate for storing the bales.

The steps involving people productivity will have upper limits, but they have not been established yet. This is a good point to remind you that making product as fast as possible is not the goal. Making product right the first time at a sustainable rate is the true goal. Time spent correcting mistakes is a part of the process too. Only this waste should be eliminated. When an operator works too fast, mistakes are likely to occur. In addition, the rate probably will drop off as the operator becomes tired. Monitoring the defect rate should pick up this problem.

None of the four inputs is able to generate a production rate higher than the baler capacity, so unless the baler stops operating properly, each input stage is its own bottleneck and defines the process capacity. For fully loaded trucks, the process capacity is 25 bales an hour; it is closer to 20 bales an hour for partially loaded trucks. Transit vans are only capable of delivering 9 bales

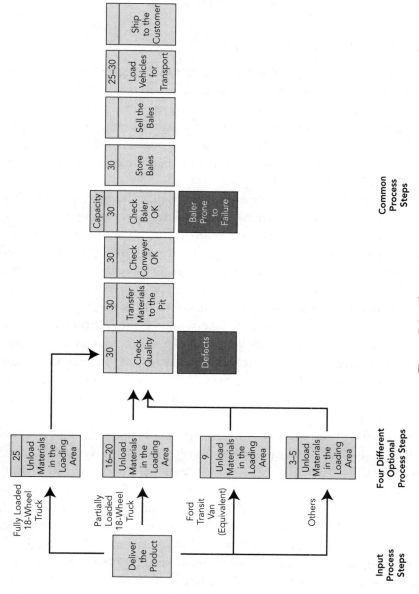

Figure 5.3 The capacity map.

an hour. This is an issue owing to the number of loads that are delivered in this way. The fourth line, designed for small volumes, can be as low as 3 bales per hour. This is a special case and does not happen very often.

Improving the Capacity

To begin making improvements, we have to investigate what limits the production rates at every stage. The best, consistent rate of 25 bales an hour is achieved only when a full load is delivered. This is running at an acceptable level for the moment. Should we want to improve on this, we would need to speed up rate the trucks pass through the unloading process. This would be a good improvement to make because it also would benefit the others. But it is not the best place to start.

Eighteen wheelers tend to be fully loaded but can be only partially full—from 60 to 80 percent. The transit-sized "white" vans coming from small, local suppliers tend to carry only a quarter of the quantity of an eighteen wheeler. Fortunately, owing to a faster turnaround, they produce a better production rate than the load would suggest. Because they carried such a large proportion of the deliveries, the vans were prioritized for improvement.

Owing to the size and maneuverability of the vans, they can be weighed faster and are able to negotiate the load/unload routes quicker. Even so, a *drop* in capacity of 64 percent is too much of a loss to be acceptable. When a van is unloaded, output is reduced by 16 bales an hour. To quantify the loss in income (assuming that all units made can be sold), Figure 5.4 illustrates the impact of lost production on income up to the maximum rate of 25 bales in an hour.

At a selling price of $15 (£10) per bale, unloading white vans loses 16 bales or $240 (£160) in one hour. This is $1,920 (£1,280) over an eight-hour shift. At $60 (£40) per bale, we lose $960 (£640) per hour or $7,680 (£5120) per shift. Considering the latter example only, for a five-shift week and 50 weeks in a year, if we *only* unloaded white vans, we would be losing $1,920,000 (£1,280,000). In reality, we need to know how many white vans we do unload in a year. It could be as high as 25 percent of the throughput. If 25 percent is an accurate rate, we are losing around $480,000 (£320,000). If the bales sold for only $30 (£20) each, the loss would be half this value ($240,000 or £160,000), which is still not to be scoffed at. With accurate data, you can make a better estimate.

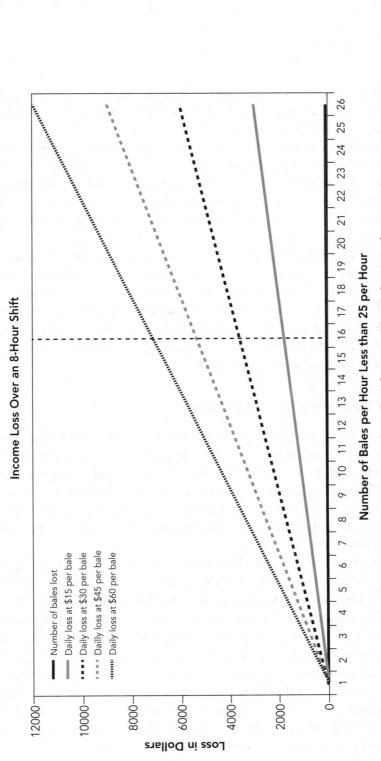

Figure 5.4 Sales losses over an eight-hour shift—up to 25 units lost in an hour.

How did the team make improvements? The capacity map made it obvious where the fixes were needed, but it was up to the team to find the solution. What do you think we did?

It was discovered that eighteen wheelers were always waiting to unload materials, and we know that *waiting* is a waste. So how can we eliminate or reduce the waiting? The ideal solution would be to only unload full trucks and ignore the vans, but we would be losing a lot of business. What is the problem? Trucks can't get access while a van is unloading. How do we get rid of the vans?

The team brainstormed for ideas. The solution became obvious. Once the capacity issue was recognized and the cause discovered, the solution was to create a second bay for unloading white vans. The bay was designed as a simple unload area—no frills. When a suitable amount was reached, the material was fed to the drum for baling while an empty eighteen wheeler was replaced by the next full truck. Further improvements could be a second drum and conveyor to feed the baler.

The Cabinet Manufacturer

This is a more complex process. There are two lines. The first makes the doors; the second makes the sides, the cabinet bodies, and the internal shelving. The doors can be intricate, with a computer numeric control (CNC) machine ensuring precision. Optional finishes are available, including paint, varnish, and veneers. The final assembly and hardware attachment are done by hand. Equipment has capacities but is often limited by the processes.

So where do we start? The capacity map in Figure 5.5 is an approximation of the process. It does not include all the steps but does contain all the key production rates. Ranges have been included where outputs are variable. The data have been taken from vendor manuals or determined by discussions with the production manager and operators. The discussion also provides a brief summary of the main issues.

This map shows the variation in capacities across the processes. The sanding process (on the calibration sander) has a wide range of throughput. Fitting veneers is a very unreliable process, ranging from 5 to 20 units in an hour. Drying the units after lacquer or painting takes a long time and is limited by space (the units are normally allowed to dry overnight).

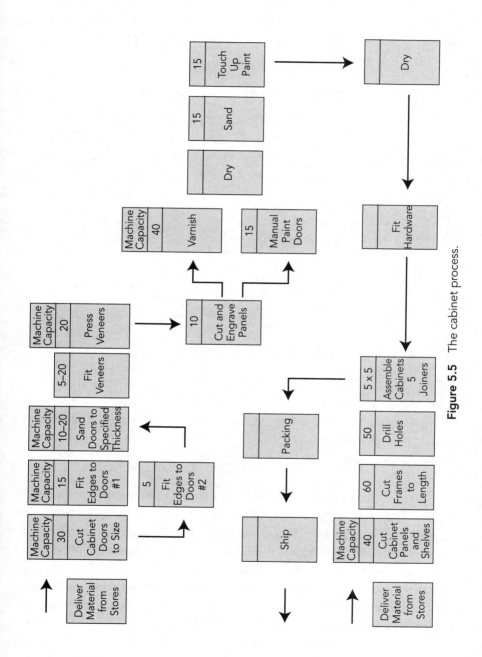

Figure 5.5 The cabinet process.

A spreadsheet and a bar graph are also created. Refer to Table 5.1 and Figure 5.6. The spreadsheet helps us see the time taken to produce one unit and the maximum throughput possible for one hour, one shift, and one week.

Table 5.1 Capacity Data Analysis Table

		Number of Units (Throughput)		
Process Stage	Time/Job Minutes	Hour	Day (8 hours)	1 Week (40 Hours)
Cut doors to size—beam saw	2.0	30	240	1200
Edge bander	4.0	15	120	600
Calibration sander	3.0	20	160	800
Fit veneers	3.0	20	160	800
Press veneers	3.0	20	160	800
CNC (average)	6.0	10	80	400
Varnish (lacquer)	1.3	45	360	1800
Paint	4.0	15	120	600
Dry			0	0
Sanding	4.0	15	120	600
Touch up	4.0	15	120	600
Dry			0	0
Cabinet panels and shelves	2.0	30	240	1200
Frames cut to length	1.0	60	480	2400
Drill mounting holes	1.2	50	400	2000
Joiners assemble (per man)	12.0	5	40	200

The bar chart compares all the throughputs. The higher the bar, the greater is the throughput. We can immediately see that the smallest bar is the limiting process—the CNC tool. It can only produce 10 units in an hour. The assembly rate for the cabinets is lower, at 5 units per hour, but we have 5 joiners to do the work, bringing the throughput to 25 units, as illustrated by the dotted rectangle. The other two rectangles on the chart represent steps where a range of outputs is produced. It is rare for either of these process steps to achieve their maximum rates. The team will need to establish the reasons.

Team analysis found that the calibration sander was not always being used properly. The operators often took too much material off in one cut,

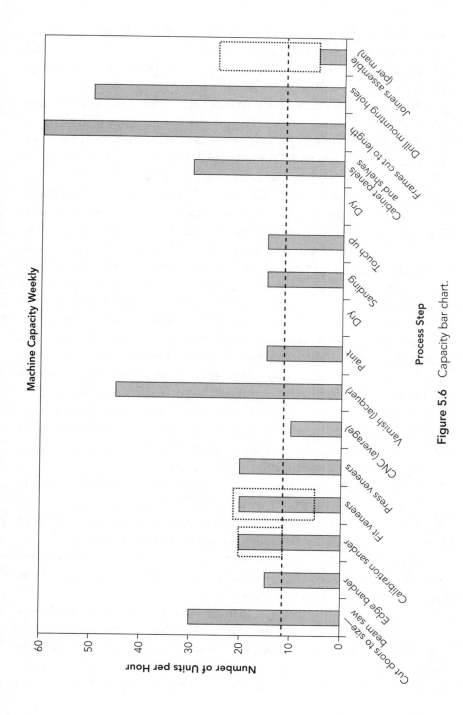

Figure 5.6 Capacity bar chart.

which could tilt the sanding belt. To compensate for the slope, the operator would feed the panels through correctly and then rotate the panels through 180 degrees to resand the high side. This accounts for the step in the range; that is, it was either a throughput of 10 or a throughput of 20.

The veneer issue is interesting. Where variation in throughput is seen, there is usually a problem to fix. The operator was provided with a sheet to record her hourly production rate over a day. She started at 20, and then her rate dropped over a couple of hours to 5. On a couple of occasions, she achieved 0. The cause: She could not find the veneers. The solution: The warehouse person set out the following day's work in advance. Improving the layout of the complete veneer process increased the throughput here. There also were experiments with the adhesives to reduce the pressing time.

The CNC throughput was more difficult to resolve. Improvements included offline programming so that the operator could spend time running the tool and less setting it up, training a backup operator to enable covering of breaks, introducing a late shift for the CNC machine, some jobs being carried out manually, and long-term plans to buy another machine.

Therefore, using the capacity map, we want to know what it is that stops us from reaching the maximum possible production rate and to find a way to reduce any resistance to flow. There are a few established improvement methods: doing a job in parallel so that it does not affect production, using two operators in place of one, making sure that the operator does not have to wait for materials, having components delivered to the tool rather than having the operator get them, making sure that the operator is trained, and making sure that the machine operates properly.

When a company becomes experienced enough, it can add OEE data to the capacity chart. OEE refers to equipment. It is basically the percentage of production achieved as a fraction of the capacity of the machine when everything works perfectly. If the capacity of a machine is 100 units in a month but we get only 50 units of *sellable* quality, the OEE is 50 percent. The reasons for the reduction in performance are the same as we look for in this book—quality, performance, and availability.

When we create a process map, we ask the operators what their maximum throughput is. It is likely to be lower than the rate the machine is capable of. Precise data could be found from the equipment manuals or the manufacturer of the equipment, but we don't want to scare the operators. The process of improvement is not about cutting jobs, although this is always a fear of the operators. Consequently, operators will have a tendency to underestimate. Initially, I would accept the number they claim. Then, as they begin to learn the process, eliminate the issues, and develop trust in the application of Lean, we can start to increase the capacity.

Please refer to Chapter 6 for a variation on the value-stream map that considers capacity in terms of partial value.

CHAPTER 6

Lean Manufacturing, the Value-Stream Map, and Partial Value

I have tried to write this book along the lines of a conversation—to introduce the relevant themes as they arose naturally. If you started reading at the beginning of this book, by now you will have a pretty good appreciation of Lean, but the trainer in me believes in reinforcing—so it is time for a quick refresher. I thought I would go over the key principles with a slightly different slant this time and provide some different examples.

A Comment on Mass Production

In the widest sense, mass production is the opposite of Lean Manufacturing. Mass production evolved to make as much of a product as possible at a time when the demand for the product was such that there could never be too many to sell.

Every time I think I know when mass production started, I find that there was something earlier. First, I thought it started with Henry Ford. Then I discovered that the techniques he used were derived in England a hundred years earlier. It was then that Richard Arkwright set up his first textile factory in the Midlands. Arkwright used water power to drive belts and pulleys that ran multiple machines, and he used semiskilled operators. The efficiency of his systems produced enormous amounts of materials and made him oodles of money. To maintain his manufacturing advantage, the water-powered cotton mill process was well protected. Because of determination to keep the knowledge in the United Kingdom, cotton engineers basically were banned from emigrating by the government. But what a surprise! The plan failed. The technique spread eventually to the United States, where the cotton industry grew by leaps and bounds. Samuel Slater, an apprentice of Arkwright's partner, has been credited with taking the methodology to America. His methods were adopted by many companies, one of which was the Springfield Rifle Company—the company on which Henry based his factory. Slater eventually became known (in the United States) as the "founder of the industrial revolution."

Clearly, I now knew where mass production started—until I discovered that in Venice, hundreds of years earlier, a sailing ship could be built in less than six weeks. As it happens, they were being mass produced—in shipbuilding terms. The wood was precut into shapes, and a number was marked on each piece. This allowed semiskilled workers to assemble the numbered parts following a simple plan. This reminded me of the people who write numbers on the backs of jigsaw pieces to make assembly easier.

It was variation in part sizes that limited true mass production. Manufactured parts were all different sizes. We had to wait until advances in measurement accuracy made machined parts identical and fully interchangeable. Guns, for example, were made one at a time. Individual parts were tweaked to fit together until Joseph Whitworth, an engineer, invented an accurate measuring device that enabled parts to be built to a specific predetermined size. Then, if a manufactured part fell within an upper and a lower tolerance, it could be used in the assembly of any of a number of identical items. Hundreds of parts could be made at one time and be assembled later with complete confidence that they would fit. The Springfield Rifle Company made accurate parts and assembled them on a production line. It was such an efficient system that Henry Ford based his first factory on the method.

In the early days of mass production, there were so many people wanting to buy the products that production delays were all but irrelevant: People were prepared to wait. In the car industry, so many of the same parts were made that groups of the same tools would cluster together in "production villages," all making thousands of identical parts. When completed, the batches of parts then would move to storage areas and sit there, waiting for the operators from the next "village" of tools to collect the parts and do their bit to them.

I sometimes get told by companies that their product is not made on a production line, so Lean will not work for them. This is my fault. It is easier to explain efficiencies in terms of continuous production. But Lean techniques do work for everyone—from mass producers to single-piece engineering jobbing shops. They all have a need to find areas of waste and control costs. Even mass producers embrace Lean—they, too, now manufacture in smaller quantities (batches). This gives them better control of lead times when making multiple products and reduces the unnecessary cost of bulk storage. Their production rates are still linked directly to customer demand.

Why would companies that make products so popular or essential that they always sell ever want to use Lean? For manufactures of drinks, clothing, screws, paper clips, coat hangers, aspirin, toilet papers, bottles, light bulbs, and so on, mass production is the way. However, even these processes suffer waste. Inefficient processes still can kill what would otherwise be a production gold mine. If products are too expensive to make, canny rivals always will find ways to undercut the price. Global competition is a way of life, with new, low-cost economies making parts for fractions of what it costs to make them in the West. Corporations and enterprising companies continually seek ways to outsource manufacture and set up shop in the areas where costs are lower. They are simply trying to remain competitive and make a profit.

Continuous Improvement

One of the objectives of this book is to help you to use Lean theory and achieve its benefits. Not everyone agrees with Lean principles. I have spoken to managing directors and managers who believe that Lean has no value—that it adds nothing to the bottom line. This is not my experience. I am basing my opinion on the 10+ years I have been applying the theories and the years before that when I used the same techniques in different guises. During the early years, much of the Lean I used was derived from other improvement processes and some good, old-fashioned common sense.

A library needs to have more than one book. Thus, although Lean works well as a stand-alone process, I have found that a company will get better results if Lean is used in concert with other techniques. Continuous-improvement methodologies tend to be based on similar frameworks, their differentiators being a few specialized steps embedded within them. My intention is to teach Lean but also to show you where I have found it useful to include the supplements. My favorite add-ins are from total productive maintenance (TPM), reliability centered maintenance (RCM), and the theory of constraints (TOC), single-minute exchange of die (SMED), overall equipment efficiency (OEE), 5S, and capacity flows. RCM is more theoretical in operation but no less practical in its outcomes; it is a kind of upmarket failure modes and effects analysis (FMEA). Developed for the aircraft industry, it anticipates what *can* go wrong and defines the consequences of a failure. This fits nicely into production equipment and its operators. By using simple accounting and spreadsheets, one can quantify losses and compare their impact, as promoted in RCM.

On its own, Lean Manufacturing provides a huge number of benefits. It is important that you know how Lean works and what tools are available. However, unlike companies that only promote their own products, I would like you to become more like an independent financial advisor and recommend what is best for the company. If there is a better way to carry out a job, then do it that way.

Lean, in common with other improvement processes, creates extra time to make products, or to be more precise, it recovers wasted time. The following statement should be obvious, but it would surprise you how few people recognize it for what it is: If you do a job correctly the first time, you will not need to waste any time having to put right any mistakes. In addition to the saved personnel-hours, you also save on the materials and any facilities (e.g., light, heat, electricity, water, compressed air, etc.) needed to carry out the work.

But . . . the time saved becomes a benefit *only* if you use it positively. If overtime was paid to complete the repeat work, you will have saved money. There is little gain in making one step 10 minutes shorter if the operator still has to stand around and wait for the next job to arrive. Use recovered time to make more *sellable* product; set up small teams to improve other processes or design new products.

Lean Manufacturing

The first point to appreciate in Lean is that it focuses on the customer. Satisfying the needs of the customer is paramount; the customer is the one who keeps your business alive. You should always know what the customer wants from your product or service. Why does he buy from you rather than another? The five key principles of Lean are designed to identify the following:

1. What does the customer *value*?
2. How does product value flow through the process—the *value stream*?
3. How can you manufacture your product more effectively so that it *flows* through each step and avoids obstructions?
4. How can you make your product efficiently in smaller quantities and still save money? Can you make only what is needed and supply at the *pull* of the customer?
5. Strive for *perfection*.

Consider each principle in order. After reading the first five chapters, you already should have a good appreciation of them. It is important that these concepts become second nature to the Lean facilitator, so this is the final set of examples I will give.

Value

Value is an elusive quality. It is like *worth*. Advertisements like to proclaim, "Buy one of these and get a free gift worth $29.99." Worth it to whom? A glass of water has no value to a drowning person—but what about a person lost in a desert? At an auction, people will bid to a high price if they feel that the item is worth it to them. Its value might be related to the item's age; it could have been created by a popular artist; it could depend on how well the item is made; or it might add to a personal collection or have a potentially high resale value. Whatever—the object must have something that the buyer sees as value.

Why buy a particular car? What does the purchaser value in it? Is it the speed, the style, the make, the fuel economy, the number of seats, the safety features, the sun roof, the accessories, the color, the license plate, or the price? It even could be the quality of service the seller provides. There will be other criteria, some of which you probably would be surprised to admit. Perceived value is what makes one person buy something another would not. It is the same argument when you buy a pair of high-priced designer running shoes as opposed to an equivalent lower-cost pair. In eight years, I have been in only one training class where someone took offense to this statement. He was a runner and went into great detail on the real benefits of the shoes. I felt like Bugs Bunny after being whacked on the head with a plank. In the case of four-wheel-drive cars, how many of them go further "off road" than a supermarket parking lot? In the case of paintings, how can one be worth $50 million as opposed to just the cost of the paint, the canvas, and the artist's time? Owning the items might give the purchaser a degree of satisfaction, but the value also might be in others knowing that you own it. *Value* is why some people will pay for designer clothes or even poor-quality copies—the value is in the label, which explains why labels have moved from the inside of the garments to the outside.

Mobile phones are another area where perceived value has a high influence. I rarely use the camera or video features of my phone, but millions do. Perhaps I am a photo snob, but I prefer to use a "real" camera for the quality. Notice where the value lies for me: It is a real camera! I would hardly ever search the Web on a phone unless I absolutely needed to. I find the phone screen too small to enjoy reading from it. Today's phones include satellite navigation, e-mail, video, word processing, and spreadsheets (well, four cells of one!) and have enough memory to hold my entire music collection. I am continually surprised at how many texts people want to send. A recent advertisement was offering 5,000 free texts in a month. This is 161 texts each day of a 31-day month. Somehow 5,000 seems more than unlimited. So it begs the question: Why should *I* buy a particular phone? What are the features that *I* see as having value? What do *I* want the phone to do for me?

The same considerations will be made of your product. Why would I buy from you? If yours is a unique product, there is a massive advantage. Usually, there is a close alternative, a similar product, and yet I still buy the one made by your company. What gives your product more value to me? Maybe I prefer the service or level of friendship. People tend to buy from other people, not companies. There are shops that I will never use because I don't like the way they operate.

Is delivery reliability an important feature? I might not need a same- or next-day delivery, but I would insist that it arrives when promised. Do you have lots of issues that push up the manufacturing cost—and consequently the price? Are your processes very inefficient with regularly late deliveries? Why would I want to pay for your mistakes? If a plumber floods my kitchen by cutting the wrong pipe, should I pay for his time to put it right?

The value of your product will vary from customer to customer. As a manufacturer, how do I decide what features to add to my product? What will increase the product's perceived value? On what should I spend time and money developing for future products? How do I find out what the customer values? The answer is not too difficult: Ask!

The Value Stream

I will not discuss this principle in too much detail because it was covered in Chapter 3. There is another variation of the process map, shown in Figure 6.1a. It is called the *value-stream map* (VSM). I tend to prefer the process map and use it to find value. I prefer to use red Post-it Notes to define losses. I think that they have more visual impact, and I prefer them to be in the correct positions on the process.

In the VSM, just as in a process map, every step is laid out sequentially. Each step is reviewed individually to see if it adds value to the product or not. In any manufacturing process, the object is for every step to add value. The map looks for issues to resolve, that is, the steps that do not add value. Non-value-adding steps, like problems, cost time and money. If you can find out where they are and what they do, you can fix them.

VSMs make it easy to add up the time taken by value-adding and non-value-adding steps and give an immediate overview of the efficiency of the total time taken to make the product. By adding the steps that add value, you know how long the process *should* take. The time it actually takes includes all the non-value-adding steps, too. In Figure 6.1a, you have an estimated total of non-value-adding time of 55 minutes out of a total of 135 minutes. You have to exclude the time to sell the bales and the time to ship because we do not know what the values are. Ideally, the bales should have a quick turnover or you are just adding to inventory as a loss.

Steps that add value seem clear. They are the logic 1 steps in Figure 6.1a. Yet I think that perceiving value is not always a straight "Yes" or "No" option. There are steps that you would agree add value, but only *some* value. These steps are the inefficient ones. If you only see them as adding value, why would you ever try to improve them? In my opinion, the value-adding steps in Figure 6.1a are driving the van to the unloading area, unloading the materials, transferring the materials to the drum, selling the product (this is a tricky one as regards value—it is an administrative task), loading the product, and shipping the product to the customer. However, even these steps can be improved. Is it possible to reduce the drive time to the unload area, for example? Is it possible to unload directly to the drum and eliminate a bit of double handling?

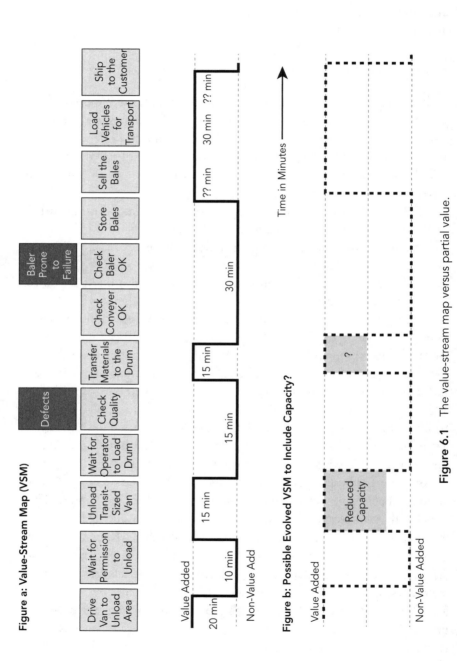

Figure 6.1 The value-stream map versus partial value.

The non-value-adding steps include both the "waiting" steps. You never should have to wait for permission to unload. This should be a given by the time you reach the unload area. Neither should you need to wait for an operator. Checking the quality is essential, although it is a waste. Checking the conveyor and baler need only be a confirming glance. Preventive maintenance should have minimized any risk, so proper operation should be expected.

Less obvious, but you know of it because of the capacity map, is that unloading a transit van adds only 33 percent of the value of unloading a truck. It is value-adding, but you need to unload three vans to get the same quantity of value as unloading one truck. I think that you need an intermediate-value step. The time it takes a white van to unload 30 bales is three times longer than the 15 minutes shown as added value. Should you not, then, have an extra 30 minutes of non-value-adding time to represent the slower speed? The same argument would hold true for a machine running at 50 percent speed: Should you have an equivalent non-value-adding time?

In addition to the capacity map, capacity table, and capacity bar chart, maybe you can add a *value- and capacity-stream map*, as shown in Figure 6.1b. In this new map, the unload step includes a hatched area representing a two-thirds reduction in value. The step that transfers material to the drum must have some amount of waste built in. It is mostly double handling, and any job that is carried out twice is a waste. So I have represented this as a hatched area with a question mark, suggesting there is an unknown amount of potential savings, but it is definitely worthy of analysis.

When I work with companies or teams, I am also working with people. This must never be forgotten. So I must highlight the positives, too, to avoid alienating my customers. Where an operator or company does something particularly well, I make sure that it is recognized and appreciated. Ask whether the same procedure might be used elsewhere.

Flow

Mass production does not ignore flow. Where products are made in very large numbers, such as bottles, drinks, or food, the systems tend to become automated. In the case of drinks, empty bottles are delivered in pallets to the depalletizer, where they are unloaded. Then the beauty of the flow begins.

The bottles are fed through washers and a whole series of automatic processes as they glide along lines of conveyor belts. Each conveyor needs to have its speed set to enable the *flow* of materials to move gracefully from belt to belt. Where the moving belt interfaces with a piece of equipment, the flow must be synchronized to allow the machine to take bottles when it wants them. If the incoming bottles move too fast, there will be a buildup at the entry point to the machine that will cause it to jam, divert the bottles into a holding area, or stop the belt until the buildup is reduced.

Getting the flow right will enable a single bottle to run the production gauntlet without being damaged or delayed. The capacity map also helps here. Each process step will have an operating speed set by the engineers. The speed set may not be the speed at which the machine was designed to operate. To ensure that the flow is continuous, engineers often slow the machine down when mechanical issues arise. To get the overall flow right, the *slowest* production rate defines the speed of the line—its heartbeat. Thus the speed of the line ultimately is defined by the bottleneck. When you apply flow to manual production lines, you are looking for the same objective. You want the bottle to be passed down the line without damage or delay.

To ensure continuity, flow also has to control batch sizes. Consider a three-step process with a single operator at each step. If it takes the operator one minute to make one item and the batch size is 10, this means that operator 1 needs to make 10 parts before she passes her batch to operator 2. If you start a timer as the first operation begins, it would take 10 minutes before any product appears at the second operator. Operator 2 has just won an extra 10-minute break.

Operator 1 now starts on batch 2 while operator 2 progresses through batch 1. Operator 3, meanwhile, is having a second cup of coffee. It takes a full 20 minutes before batch 1 reaches him and all three operators are finally working.

After a further 10 minutes, the first batch leaves the production line. The complete process took 30 minutes. After this point, product will leave the line in bunches of 10 units. If the process is well set up, after 30 minutes, the product will be flowing along the line, and everyone will be working. But what if you wanted only one item for a customer? Making in batches of 10 would mean having to wait for 30 minutes to get it, and nine other parts would be put into the warehouse, waiting to be sold.

For the best flow, Lean targets a batch size of one. This would mean that step 2 would start after only one minute and step 3 after two minutes. The first complete unit would appear after three minutes, not 30 minutes—27 minutes sooner. Nothing would need to be put into the warehouse. In fact, what would be the need to keep any finished goods in the warehouse at all when you can make one in just three minutes? To add a further degree of complexity, the process would run at a *Takt* time of one minute. That is, one part comes off the line every minute. If the *Takt* time were five seconds, every time one of the seven dwarfs sang a line of "Hi ho, hi ho, it's off to work we go," a unit of product would appear at the output. (I'll bet that you are singing in your mind right now!)

Flow has no units. There will be a production rate, but it says nothing about the stops and starts and reworks. The process map, the capacity map, and the "7 wastes" would be used to find out what interrupts the flow.

There are also benefits to be achieved by companies that make complex assemblies. It is something you might overlook when training in Lean. People tend to talk in flow rates to explain benefits, but even if you are making a one-off product, you still have waste. Eliminating the waste will speed up the lead time of the individual modules. Some companies find it hard to see two components of different sizes as being essentially "the same." Consider a piston, for example: A 10-cm-diameter piston with a 30-cm movement is made in virtually the same way as a 15-cm piston with a 45-cm movement. A design engineer would disagree because he or she knows the complexities. I would be looking for a link between the times to make them based on the similarities. By the way, I just made up the piston sizes to make a point.

Pull

Pull is an easier concept. If you want a suit of clothes or a pair of shoes made to measure, you visit the manufacturer, pick the material and style, get measured, and he or she makes the item as you want it. The items are being made at the *pull* of the customer. They are not being made first and sold to you later, which is known as *push*.

If the product is beer, rather than make tens of thousands of cans of each type, you try to predict what the customer wants and make multiples of that amount. In this case, you are making to stock based on a forecast.

Usually, there is a buffer stock of around four to six weeks' worth of product. Making items in mass production, you run the risk of them not selling and having to be scrapped or sold off at a reduced price. In the preceding pull example, the suit or shoes have been sold. In normal circumstances, they never can be left on the rack.

In Lean Manufacturing, you try to limit how many items you make. The closer you can predict the true customer demand, the better. If goods must sit in a warehouse, you are paying for floor space. Should the items become obsolete, you have wasted all the labor, material, and overhead costs. By controlling how many you make, you reduce the number likely to end up on the shelf. In push, you make oodles of something, store the goods, and then look for someone to buy them. I find real pull harder to achieve.

Perfection

It is not unusual to find perfection as an improvement goal in Japanese continuous-improvement techniques. It is a good goal—if not the ultimate goal. With perfection as a target, you are saying that you will never stop trying to improve. As an advocate of TPM, a productivity system with zero fails as a principle; I feel that I am softening in my approach.

I have worked with so many companies that would love to fix everything they know is wrong but believe that the cost prohibits it. In reality (or perhaps it is just my failing), I concede that there comes a point where the cost to fix an issue significantly outweighs the benefit to the company. To this end, I promote quantifying the impact of a problem and displaying the amount on a bar chart or a Pareto chart. The purpose of the graph is to see what the biggest issues are.

All else being equal, you should fix the worst issues first. What do you do, then, if the worst issue will be difficult to fix or take a long time? You can consider the resolution of a few lesser problems, more easily fixed, if the total savings over the time taken make it worthwhile. Sometimes a problem must be resolved irrespective of the cost. Safety would be one example of this, as would be improving employee morale. In any case, where processes are analyzed continually, there always will be a bigger issue that needs to be fixed or improved.

The "7 Wastes"

By now you should be fed up reading about them. If you learn nothing else, learn the "7 wastes." The "7 wastes" are the basic toolkit to be used in improving the performance of a factory. Couple this approach with a few diagnostic techniques and some problem-solving skills, and you are well on the way to making significant positive changes.

There are purists, many of them very good, who believe that Lean must be applied in the order of the preceding principles: value, value stream, flow, pull, and perfection. It may be the case that such purists are right. My experience is a bit different. I have spent much of my career fixing problems. I also have had time constraints with companies. So let's look at the issue from the perspective of fixing as much as we can in the shortest time possible. Using a jigsaw analogy, we can still apply all the parts but possibly be a little flexible with the order of assembly.

What is the first thing you need to do to fix a problem? Diagnose it. You need to know that the problem exists. This brings us back to mapping. Laying down the process as a visual diagram gives you a chance to see what happens. Adding the initial red Post-its helps you to see the problems that you *know* exist. Some of them may have been points added to the "parking lot" when the map was being created. The "7 wastes" are the problems that anyone can suffer at any process step. Perhaps you don't even recognize them as being problems—they have always been there. Maybe you do know that they exist but choose not to do anything about them. Often, the real barrier to fixing them is that you don't realize how much they are actually costing you in time, materials, and money.

I could never recall the wastes until I was introduced to TIM WOOD. This is the best acronym I have seen. I first heard it from the Manchester Manufacturing Advisory Service. The "7 wastes" involve:

1. *T*ransporting
2. *I*nventory
3. *M*ovement
4. *W*aiting
5. *O*verproduction
6. *O*verprocessing
7. *D*efects

Transporting

In any process, the goal should be to keep the distance between each process step as short as possible. Imagine a scenario, usually from an old cowboy movie, where the saloon catches fire. Folks don't run from the stream to the saloon with the buckets of water; they form a chain and pass the buckets from person to person. If there are enough people, they do it in the opposite direction too with the empty buckets. Perfect flow—and yet this was before the days of Lean. The reason they did it was because it was the fastest way to get the water to the fire.

Today, we have forklifts and trolleys. Moving material is less of a problem. We can move lots of stuff at one time and sit it on shelves, waiting to be used. This creates two problems: Having inventory sitting on shelves costs money, as does carrying materials for long distances. We need to take care when locating storage areas. They should not be placed just anywhere.

Transporting materials takes time. The problem is magnified if it is the production operator who has to move them. A two-minute walk from one tool to another at opposite ends of the factory is not too much if it only happens once, but if it happens regularly, the time mounts up. I know that I said that it was only a two-minute trip, but it never takes the minimum time. Billy will bump into Fred, Pete, or Jody and pass on a few pleasantries. Worse still, the operator is not making any product during the deliveries.

How much time are you losing? What is the loss in personnel-hours? How much product could have been made in the time lost? Convert the figures to money. Scale the loss up to a 12-month period. Quantifying the time in terms of money will motivate the team to find a solution. The easiest way is usually to keep the workstations, materials, and tools as close together as possible. Try using a spaghetti diagram to see how much walking or transport is required by the process.

Inventory

Inventory includes raw materials and finished goods. Every item costs the company money. As material passes through the process, value is added at every step, with the maximum value being achieved when the item has been completed.

Inventory has a selling price. This is where the profit is derived. It has a virtual cost generated by the number of personnel-hours used to make it. It also has a material cost that will be much lower. During manufacture, there also will be a proportional cost based on company overheads—equipment, power, buildings, heating, phones, indirect labor, and administration. Inventory ties up money. You also have an option to purchase materials or subassemblies in bulk from a cheaper country. If you decide to do so, ensure the following:

- That paying for them will not cause cash-flow issues
- That if you borrow money to pay for them, you have factored in the interest charges
- That the goods will sell
- That you don't fill up your warehouse and need to pay extra for storage

If possible, think of inventory as perishable, and you don't want it to go bad before it can be used. Goods in storage can be lost, damaged, become obsolete, be used by accident, or need to be scrapped. I am not a just-in-time purist, but then I am not a fully fledged just-in-case person either.

Systems have to be developed to ensure that a supplier can be relied on to meet his deliveries and have an early warning system to flag issues in advance. Think about a risk assessment for late delivery from your supplier. If the goods don't arrive, what is the backup plan? How long will it take to get materials from somewhere else? Is there an alternative supplier? What will your customer do to help you if her order is late? Are there any penalties for late delivery?

Always consider quality issues from your supplier.

Movement

This is always my smallest section. Movement, as I see it, is like ergonomics. The operator's workstation should be arranged to make doing the task as easy as possible. Avoid unnecessary bending, stretching, twisting, or walking. All tools should be at hand. Any component parts should be easy to reach, and it would be nice if the containers told you when they need to be topped up (*Kanban*).

The best person to set up the operator's workstation is the operator. Use a process map. What are the issues operators find when they do their jobs? Do they have the tools they need, or do they have to share? Are tools and materials at arm's length? Are the tools used most nearest to where the operator stands? If the operator uses instructions, where are the instructions located? 5S methodology is invaluable when setting up a work area.

Waiting

Waiting is one of the biggest offenders in the "7 wastes" palette. Waiting happens all the time and always with a good excuse. In fact, it is usually the same excuse—time and time again. What do we wait for?

- Operators
- Materials and parts (including product coming from previous steps)
- Instructions (dimensions, drawings, which job, which process, how to operate the tool)
- Equipment (performance, availability)
- Tools
- Forklifts
- Quality checks

Eliminating waiting requires advanced planning and organization. If you view waiting time like the meter in a taxi, racking up the cost, you become much more aware when the taxi isn't moving. This brings us back to quantifying costs. How much time is being wasted at each process step? What is the time in personnel-hours? How much product could have been made in this time? Use the number to prioritize which issues to fix first.

I was once very impressed by a production engineer showing me the area for which he was responsible. There were some very complex presses that needed to be changed over every now and again. "I can do a changeover in 20 minutes," he said, "but sometimes I have to wait an hour for the forklift to arrive!"

Overproduction

This is the mass production versus Lean manufacturing debate. Mass production looks for economies of scale and produces in large batch sizes. Lean strives for a batch size of one. Occasionally, a company has to decide on its operating model. Is it necessary to make to stock because the customers *expect* to be able to phone and get what they want immediately? Many of these customers tend to be service companies or agents.

Computers for sale in large retail superstores tend not to be the most modern. They are built to a selling price, with no specific customer in mind other than the store that will sell them. The buyers use sales statistics to estimate how many units they will need to buy to bring their stock back up to the desired level. Unsold products are likely to become the special offers when sales come.

State-of-the-art PCs are built using *pull*—after the order is placed by the customer. Even then, they are not usually designed completely from scratch but are based on several approved options normally chosen by the company because it knows the options will work in its product.

Think, then, of the PC superstores, where the units are not sold at a planned rate. If you make or buy too many units, you know that they will need to be stored somewhere, so they will need space (which comes at a premium) and heating. Selling them also could take discounts, extras being offered, or modifications added—all at extra cost to entice someone to buy them. Cars had the same issues, and for many companies, this is still not perfect. But now, thanks to Toyota and the *Toyota Production System* (aka *Lean Manufacturing*), car companies are controlling the number of the models they make and designing their processes to satisfy this need.

Making to order is the goal of Lean (pull). If this is not possible, sales forecasting (otherwise known as guessing) must be as accurate as possible. It is essential that you know what your customers want and that your product is kept up-to-date. The lead time for orders should be controlled. If you do make to stock, is it possible to modularize your product so that key modules can be used in more than one product type?

The last word on batch sizes: If you have a batch size of 500 and there is a production mistake, then 500 is the potential number of failures that will be manufactured before the error is discovered. At this point, you will need to scrap or rework the lot. If the batch size had been smaller, the error would have been reduced.

If the time taken changing from one setup to another prevents you from reducing batch size, try applying the SMED technique. It is yet another productivity tool developed by Toyota.

Overprocessing

Overprocessing means having more steps than needed in a procedure. They will add to the total cost by using unnecessary materials or by increasing the time it takes to produce or to complete the process. Overprocessing was explained in Chapters 3 and 5.

The process map is where this would be considered. Each step of the process has to be reviewed to see if it is needed. This includes paperwork (information). How many times do you need to type the same data? I remember one company that had a spreadsheet for recording everything made in a month, a monthly report for putting into words the same information as the spreadsheet, and a face-to-face review where you updated the operations manager on everything that was written on the spreadsheet and the monthly report. All these data then were added to a master spreadsheet. Overprocessing is often the same as repetition of work.

I mentioned SMED previously. The whole procedure is in my first book, *Total Productive Maintenance*, as are many of the processes I have mentioned in this book. In SMED, the objective is to reduce the time of the changeover process. To help with the analysis, we need to ask a few questions about the steps in the process:

1. Why do we do this step?
2. Is the step needed?
3. Why do we do it now?
4. Why do we do it this way?
5. Is there an easier way to do it?
6. Does it need to be done now?
7. Can we prepare anything in advance: preheat it, precool it, preclean it?
8. Would it be better to do it at a different point in the sequence?
9. Can it be become an external part? (An *external part* is something that can be done off-line while the process is running.)
10. Can it be eliminated?
11. Can the time be reduced in any way?
12. Could two tasks be done at the same time?
13. Would two people doing the job make it faster?
14. Can the step be grouped with others and preassembled?

I once saved a company 17 minutes on a changeover. Not much, I hear you say. In fact, the saving was magnificent because the company carried out more than 50 changes each day. SMED is a simple and powerful process. Even so, if the 14 questions are considered, you will get a good idea about steps that lead to overprocessing.

Defects

Defects and rework are symptoms of poor quality. Everyone knows what they are. The odd thing from my perspective is that some organizations have entire production lines with the sole purpose of reworking faulty items. Why?

Once an item has been made, labor, materials, and facilities all have been used. To rework it, you need even more labor, more facilities, and usually more materials. In addition, the next stage in the process has had to wait before it can carry out its part. The waiting costs the company more labor and production losses. There even may be a need for an express delivery to the customer at extra cost. Worst case is a lost customer. Defects cost more than twice the original cost—often much more.

How do you stop defects? The first step is to establish where they are being generated. More often than not, in-house quality checks are not performed, and faulty parts are passed on to the next worker. This could be due to faulty equipment or tools. Alternatively, if the company has a "blame culture," no one will admit to a problem because it gets them into trouble. Sometimes there are no quality checks because there are no documented *standards* for the finished job.

You need some active problem solving here. Use cross-functional teams. Lack of information, often ignored in production issues, needs to be put right. Proper instructions with error limits need to be introduced. Formal quality checks must be carried out until the employees can be trusted not to pass on faulty parts to the next workstation or until operators reject faulty parts coming to them. Visual tracking should be introduced. A white board with a *big* chart can show the number of defects. This helps to keep the issue of quality as a clear objective. Another good chart is to show the equivalent cost in cash—both maintained by the operators, not the managers.

Never accept that defects and rework are normal. Question everything: The worst managers are those who muzzle their staff and refuse to implement improvements if they may make them look bad. Besides, the scrap and rework figures (if monitored) are already likely to do that.

Conclusion

Finding the wastes requires implementation of a proactive process. I recommend reviewing maps annually. Looking for problems is not a one-off project but should be the first stage of an ongoing continuous-improvement plan.

It takes time to find improvements. As solutions are introduced, time savings will be made. The system will pay for itself. A 10 percent production increase in a 40-hour week equates roughly to four personnel-hours per person involved. Find out what a problem costs and what it will cost to put it right. The calculation does not have to be perfect.

I often hear, "We tried improvement schemes before, but they didn't work." The most common reason for not working is managers not buying into the process or abandoning the process at the first hurdle. "We saved money, but the problems came back." Finding and fixing the issues are not enough. Systems must be put in place to avoid further issues. If they are not, the improvements will not be sustained.

If a company's managing directors expect Lean to be "done to them," the game is lost before it starts.

CHAPTER 7

Problem Solving and Decision Making

So far we have discussed *how* to identify the areas in your company or process that need improvement, and if you applied the big picture map, the process map, or the capacity map, you also should have a list of problems that need to be put right. Now comes the hard part: You need to fix them. I will illustrate this using some technical examples. They are easier to explain, but the *process* is the same for all problems.

The Root Cause

The thing is that, in my experience, most people solve problems by guesswork. Darters leap around trying to fix everything that comes to mind, a bit like technical historians, who only look for faults they have seen before. The best at solving problems are systematic and use mostly logic to arrive at solutions. In my experience, most people have never been shown how to analyze and resolve problems. If you ever meet an expert in fixing a particular problem, he won't be! He either fixes just the symptoms or puts duct tape on it. If the problem returns enough times for him to become an expert, it is guaranteed that he has never fixed it correctly.

I frequently refer to the root cause of problems. This is the true cause of an issue. Many people will jump to a symptom of the problem as being the cause. I had a power shower where the water pump would not switch off. The plumber explained about how complex power showers are as she replaced the pump. It made no difference. She went though all the wiring, which made no difference either. The interesting thing to me was that the pump had no electrical connection to the valve, so how did it know when to switch off. I contacted the pump supplier, who explained that the pump switched on and off based on the flow of water. It turned out that the pump was not switching off because the shower valve had a leak in it. Replacing the valve resolved the issue.

For a true root cause, the valve seal should have been replaced, but I had limited confidence in the plumber. The valve had 17 water seals inside it, and it scared me, too. Besides, there is normally a cost tradeoff in personnel-hours to fix versus the cost of a full replacement. I agreed to the replacement because I needed my shower back as soon as possible. I plan to try to replace the seals myself now that the shower is working. I suspect the same failure will happen again.

Returning to my "expert"—if any of his solutions were not to the root cause, then he failed to complete the repair. Avoiding recurrence should be a part of the solution. This requires an improved procedure, a preventive inspection, or a maintenance step to be developed. Although relevant skill is required when resolving technical issues, there are formal ways to approach problems. More likely than not, there are already people working for you now who, with a bit of support and some training, will be capable of finding real solutions to most of your chronic issues.

Evaluating the Cost of the Problems

I am stating the obvious here, but some problems will be much harder to fix than others. In such cases, although you would like to, you don't need the perfect fix immediately. This means that you might not get a 100 percent solution the first time around. Let's say that you achieve only a 50 percent improvement. If the issue costs $15,000 (£10,000) a year, you are still saving a fair amount of money. I would agree to a target of 50 percent—if you can do it quickly and it is only the first stage of the improvement operation.

Notice how quickly I introduced a cash equivalent. This helps me to get problems into perspective. Although I would prefer a 100 percent solution, I can see where a partial fix is practical, provided that I know what a 50 percent saving actually means. Having a value also helps to compare what you will get in return for the cost and effort made. It helps in prioritizing repairs.

To get a better understanding of the full impact of a problem, let us consider a company that carries out work away from its own base, say, an installation or maintenance company. If the planned job is to make a modification to a process, installing interconnecting pipe work or installing a machine, a *bill of materials* (BOM) should be printed, and all the parts needed for the job should be gathered and stored in a box for the workers to collect when needed. The expectation is that no one will touch the parts in the box. I use the word *expectation* when *hope* might be more appropriate.

When the workers turn up on site and start to do the job, it doesn't take long before they discover that parts are missing. The possible consequences are as follows:

- ▲ *The job has to be abandoned early and postponed, losing both billable hours and delaying completion of the job.* The wasted time will be generated in three areas: organizing and collecting the new parts, driving to and from the job an extra time, and there will be a loss in new billable hours because the workers still will be working on this issue.

    The lost personnel-hours cannot be recovered, extra mileage and expenses are introduced, and you now have an unhappy customer, which can be more serious than the financial costs.

    The worst-case scenario is where the missing part has to be reordered from the supplier, with the job being rescheduled. In addition to fixing the original problem, any work already completed may have to be undone to enable the process to be restored to an operational state.
- ▲ *If other work can be carried out as a stop gap, only one of the workers has to leave the job and retrieve the missing part.*
- ▲ *It is not unusual on longer jobs for missing parts to become routine.* Often the situation is caused by poor parts delivery from suppliers. The workers regularly stop what they are doing and return to the factory to retrieve spares. They don't complain; they get to go home early or start late the following day.

    Once more, we have lost personnel-hours on the job plus any time needed as the workers search for the part or stop off at the wholesaler to get a replacement.

    If a worker takes a part from another kit, we run the risk of initiating a repeat of the problem at the next customer's installation.
- ▲ *An equivalent part is installed to get the job moving again (and placate the customer).* The team has to return for a second visit to change out the part and recalibrate the system. Depending on the distance to the job,

this loses a whole day's income for a team of people, and the temporary part may have to be scrapped.

These examples are not hypothetical problems. They happen frequently despite being easy to prevent. A quick, simple fix would be the introduction of a formal procedure:

- ▲ Check the contents of the box to ensure that everything is present.
- ▲ Sign the name of the person who checked the contents, and date it.
- ▲ Seal the box shut.
- ▲ Forbid the removal of parts.
- ▲ Check that the box is still sealed before the job starts.
- ▲ Take corrective action if parts are missing.

If the seal is broken, the kit is unlikely to be complete. This procedure may not work every time. Someone will decide that her job is more important. The problems here are due to poor organization and processes. The root cause is the way jobs are planned and the informal way that the parts are controlled. Even a partial improvement will bring a positive saving for a minimum of effort.

When I wrote my first book, I spent a bit of time comparing different problem-solving methods. I found that apart from the easiest ones, such as the "5 whys," they are all pretty much the same. Each one might have a variation that makes it different, but the basic procedure is the same in all cases. In Lean, we have the "7 wastes," so if I was a trainer 20 years ago, how could I make my offering more unique? I offer the "8 wastes." The eighth waste is now universally regarded as *untapped human potential*—developing employees and increasing their existing skills. This was not included by the Japanese as a waste, I suspect, because they have always recognized the potential of their employees, and the employees have reciprocated.

In Japan, a lack of training is rare. Not so in the United Kingdom: I have been in companies where no one has been formally trained in 25 years. Employees pick up new skills by trial and error. They seem to get by, but there can be a lot of issues as they learn. Could the eighth waste simply be a natural step in the evolution of Lean in the West, where it was not normal behavior?

Kaizen or Define, Measure, Analyze, Improve, and Control (DMAIC) as a Problem-Solving Tool

As a vendor engineer, I used to travel the world fixing problems on equipment. I had a problem understanding why I could get the problem fixed when the other engineers and technicians in the factory, many of whom were far better engineers than me, could not. Eventually, I reached a conclusion: I had to get it fixed. I did not have the luxury of passing the problem on to the next shift. I *was* the next shift!

To solve a problem, it takes people, skill, ideas, and commitment. I like to follow the *Kaizen* method. *Kaizen* is a technique known for making big improvements in stages. It does not preclude normal faultfinding. Figure 7.1 shows an eight-step *Kaizen* process.

Step 1: Identify the Project or Problem to Be Solved

Recognizing something as being a problem is the first step toward fixing it. There are two ways to achieve this. You can wait until a problem appears, or you can proactively look for problems. Since the main goal of this book is diagnosis—identifying issues and opportunities using mapping—you already have learned how to find problems. If you have applied the processes already, you should have your list of issues.

From your list, however, how do you decide which problem to fix first? What makes one problem more important than another? Generally, it is the trouble the problem causes. Surely, if it involves only a bit of scrap or rework, you can live with it. It becomes important when it causes disruption to planned output or has a potential cost in lost product or sales.

Consider a taxi driver whose cab will not start. Quite simply, the driver will have no income, so he must find an immediate solution. If his problem is such that it does not force an immediate fix, something like an oil leak in the engine, the driver can keep the taxi going by topping up the oil. In effect, he can take a calculated risk. The *urgency* to repair is reduced. The risks are not insignificant: The leak can become catastrophic, the driver could forget to top up the engine, the engine could run dry because of a trip being longer than expected, or the driver could run out of oil for topping up and need to find somewhere to buy more. He still risks a dead engine, which is the

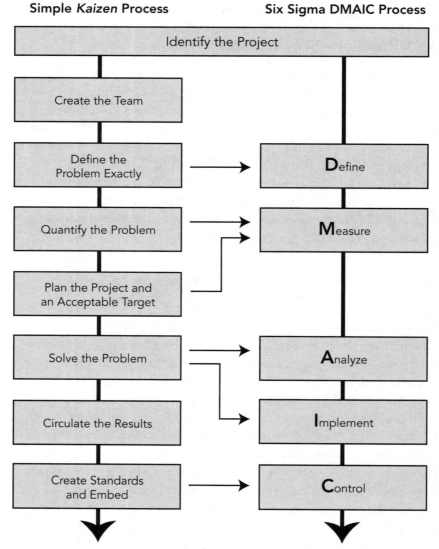

Figure 7.1 A simple *Kaizen* problem-solving flow with DMAIC comparison

same as a dead taxi; in addition, he could end up with a bigger bill in the long term and no income until the engine is repaired.

The first problems to target are those which have a direct impact on customers. The customer always should be the main priority and always should get what she wants at the quality she expects. Knowing how much an issue costs also will help to make deciding easier. (I will return to this in step 4.) A third consideration is the difficulty to fix. A difficult issue can take a lot of time. It could be wiser to use resources by fixing a number of easier issues in the same time frame. It is also possible to work on the larger issue as a longer-term, planned project.

Step 2: Create the Improvement Team

Who should work on the issue? This depends on the complexity of the problem and the options you have. If the problem is simple maintenance (or no maintenance), process issues or lack of procedures, you just need the people with equipment or process experience. Simply writing a procedure could be all that is needed to resolve the issue.

Where a small company has no in-house maintenance, the issue is slightly different. Here, an external vendor must be used for equipment knowledge. As an ex-service engineer, I have experience with companies that would call us for complex issues or personnel shortages. Often, a company uses one vendor. When I had my own company, I was asked to have a chat with a third-party maintenance vendor. The engineer and I had a completely different view of what the company needed. I advised the company to compare the cash paid to the maintenance company with the cost of having its own engineer. It was about three times more expensive to use the vendor—for less cover. The vendor, as you might expect, was nothing short of horrified. He stood to lose in excess of $75,000. As an alternative, a suitably skilled operator could be formally trained to do some basic first-line diagnosis and maintenance.

To work on a bigger problem, I like to have about four people in a cross-functional team: a process engineer or someone who knows the process, an equipment engineer or technician, an operator, and a manager. They would comprise the core team and could pull in extra support as needed. Someone from finance should be on tap to help with costing, as could someone from quality and someone from the warehouse or shipping.

I often recommend including an equipment vendor on the improvement team. In the case of using an external contractor who is not the vendor, she most probably will want to charge for her time. I also wonder whether the contractor would have enough relevant knowledge. For equipment vendors, the company is offering them a bigger opportunity—the lure of potential equipment sales or upgrades, training courses, and increased use of the vendor parts. There are more potential gains for the vendors, so free support or advice is not impossible.

Step 3: Define the Problem Exactly

This is also the first stage in the Six Sigma define, measure, analyze, improve, and control (DMAIC) process. The exception is that there are more points to consider in DMAIC.

Defining the problem is the bit most people fail to do. Converting the problem into words seems like a waste of time. Some people even say that putting the problem into words is not possible. However, doing so helps to ensure that everyone involved in the team knows precisely what the problem is. Even people who think they fully understand the issue often find that they don't when they try to put it on paper. Recording the issue also helps to define boundaries for the project. This will help to keep the team on track.

Any new issues found during the project, which is not uncommon, can be added to the list of problems for future evaluation. If a serious issue is discovered during the project or the new problem prevents a proper solution to the original problem, the work can be discussed with management, and the project scope can be expanded or changed to suit.

Step 4: Quantify the Problem

This is the same as the measure stage of DMAIC. In *Kaizen*, I am looking for more of an estimate. In DMAIC, you are looking for precision data, taking a series of accurate benchmarks and recording absolute values. The DMAIC process is very heavily data driven.

Some people like to create a list of problems, the priorities being those that they *think* are the most important, usually based on the things people complain about most often. How many would include making sure that you don't run out of washers? I have found that most people don't realize just how much any given problem is costing them, which is why so many issues are not even considered for resolution.

I get a buzz when a team of operators and engineers considers a range of issues and puts values on problems. The background chatter is usually high as the team works through the process—until we get to the total, and then there is a stone-cold silence. Team members often gasp, stare at each other, and then ask whether fixing the problems will mean a bonus. I have seen issues—previously dismissed as not important—that cost $1 million (£1 million) a year. This is why I continually drum on about working with numbers where possible. It is not something that I have done all my working life. Numbers fulfilled a need I developed as I became more involved with improvements. It is always best if the figures are precise, but a rough guide is better than no guide. If an issue is chosen subsequently as the project, the team can dig deeper into the real costs as it prepares the project justification—cost versus consequences.

I also advocate having a production log. Unless records are kept with production downtimes, much of the data will need to be averages based on estimates. If the downtime impact or frequency of recurrence is not known, try using the iteration method to come to an appropriate figure. It is easier than it sounds. It is a bit like haggling for a price in a market, except that you are negotiating with your memory and not a salesperson. A series of high and low guesses will lead progressively nearer to the real value each time you try. Most people seem to know how often it *doesn't* happen. So you start with a high number that is known to be too high.

- ▲ "Does it happen 20 times a month?" The answer would be something like, "No, it is less than that."
- ▲ "Does it happen 15 times a month?" Maybe.
- ▲ "Does it happen three times a month?" The answer could be, "No, it is more than that."
- ▲ "What about five times?"

Try 12, 7, 10, and 8 times and so on. Following the sequence in the example, nine would be close to the value. However, if there is the *slightest* chance that it is too high, consider eight or seven. When I get close but am not too certain, I always recommend underestimating. It avoids being accused of exaggerating the issue and undermining the credibility of the team.

What do you need to consider when estimating the cost? I have gone over these before, but they are well worth remembering. It is not essential to include all the following factors, but be aware that they exist. A spreadsheet with some simple formulas would greatly simplify the calculations.

1. How long is the equipment unavailable for production when the fault happens?
2. How much time was wasted waiting for a response to investigate the issue?
3. Was there any waiting for parts?
4. How much did the fault cost to fix in parts and personnel-hours?
5. Did managers have to have meetings to discuss the issues or reorganize production?
6. How many operators were affected? Remember, every operator waiting for a part from a failed step or machine still has to be paid and so is still a loss.
7. How many warehouse people had to wait?
8. What is the average hourly rate for anyone involved—operator/engineer/manager and so on, including overheads?
9. Was any overtime used (so increasing the hourly rate)?
10. How much production was lost on the main equipment and any downstream tools?
11. What is the average cost of a lost unit of product? You might want to work with a range of costs, say, low, medium, and high.
12. Was the equipment running for a longer time than normal? For example, if you needed to run a weekend shift, you introduce extra energy costs in the form of electricity, lighting, compressed air, heating, and so on.
13. How often does the issue happen in a day/week/month/year? An annual cost is best, but at a minimum, use the same time scale for all issues.
14. How much scrap or rework is there? Think materials and labor. A scrapped part is thrown away losing:
    a. The cost of all the materials used
    b. The machining/assembly time
    c. The power costs
    d. The time to physically scrap and document the part
    e. The cost to ship the scrap to the dump
    f. The cost for landfill or recycling
    g. The cost to remake the part (With all the extra time to make extra materials, perhaps a waiting time for parts to be ordered, and a potentially unhappy customer with a late delivery.)

h. A reworked part often takes more labor time to strip the part to replace or repair the damage. Additionally, there could be other consumable parts that need to be replaced because the part has been dismantled (O-rings or screws, for example). The unit then needs to be reassembled and retested. You also incur some of the scrapping costs noted earlier.

15. How much electricity, water, compressed air, or process gas has been used to remake? (The machine running costs.)
16. Did a vendor engineer have to be called?
17. Did you miss the scheduled delivery to the customer and need an overnight special delivery, air freight, or hand delivery? All at extra cost.

At a minimum, having a number of personnel-hours, a cash amount of lost production, and a cost for wasted materials and utilities tends to help in choosing the solution. This does not eliminate the executive decision on what to fix, but it might make it easier to justify any costs.

Prioritize the costs in a histogram. Try to avoid the urge to use a fancy 3D chart when a straightforward bar chart probably will be more useful. The heights of the bars can be compared more easily. Consider the top issue—the one with the highest loss in a year. It will save the most money. But is it the one to fix? In deciding, you need to consider the cost to repair, the complexity of the task, the positive impact on productivity, and the speed of resolution.

Figure 7.2 is a simple x-y chart with "Cost to Fix" as the y axis and "Impact on Productivity" as the x axis. The figure is more dramatic in color. Red equals bad, and green equals good. High cost for the fix is bad unless the impact is very, very good. Low cost to fix is good. Equally, low impact is bad, and high impact is good.

In the figure, option 1 is the best, and option 4 is the worst. The choices between option 2 and option 3 are more blurred. Thus, to reiterate the guidelines, it doesn't really matter which problem you fix first. If you fix the wrong one, fix the right one next time. Just ensure that it is an issue that *is* fixable and is not a major redesign. One problem may seem like the best option, but if you consider all the options, it may be quicker and save more cash if you tackle a few smaller issues rather than one big one. Think about

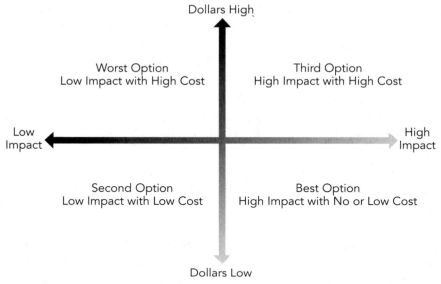

Figure 7.2 Evaluating a solution.

quality, too. Does solving the problem help the customer to get a better product? The ideal solution is:

1. Low cost, with free being best
2. High impact (a 100 percent solution)
3. A short time to fix or low personnel-hours
4. Easy to find the cause and implement the fix

Step 5: Plan the Project

This is also a part of the DMAIC measure stage. It helps to know how the systems operate before the project starts. If you have a baseline, you can compare it with the new values at the end of the project and confidently show the benefits. Sometimes, where no data are available, there is a need to collect some.

Here, you need to allocate some actual resource time for the team members to work on the issue. It is easy for me to recommend a team of three and specific engineers, but it might be too much of a commitment to tie up three people. Indeed, the company may not have engineers, and three people might be a significant percentage of the staff. Work with natural lows in the production cycle, and do the best you can to spread the load. Set targets that are achievable, but not a time so far away that momentum dissipates. The same consideration should be made for the targeted improvement. Consider my previous point: One-hundred percent is perfect, but is it practical at this point? Fifty percent is not bad *if* it is only the first step toward a better improvement and it can be achieved quickly.

The result also should be quantitative. If you had 100 fails in a week, you now have only 50. You lost 16 hours of production in a month; now you only lose six. Before the repair, you made 100 units and hour; now you can make 110. You had 20 complaints each month; now you have nine. All these savings have a cash equivalent. The team can use the savings as a measure of its success. If you have an improvements manager, she can use the figures to justify further improvements.

Step 6: Solve the Problem

This means that you solve the root cause, not the symptom. In DMAIC, solving the problem is clearly split into two sections—analyze and implement. The data were recorded in the measure stage. In the analyze stage, you are looking at the data and charts for trends, frequencies, or unusual patterns. The analyze stage also can involve statistical analyses using statistical process control (SPC) techniques. SPC is a quality tool that uses probability to warn when numbers are not performing to a *random* pattern. Don't let a fear of statistics deter you; using a program such as Minitab to analyze the data will highlight when issues arise. Alternatively, you can just follow the Western Digital rules.

Always start with the easy options first. I recommend that you never jump to a Six Sigma project as a first step. I see problem-solving tools like floors in building. The "5 whys" can be found on the ground floor, and Six Sigma lives in the penthouse. To get to the penthouse, you must pass through the other floors. Besides, to achieve a solution using Six Sigma, you still would need to use the "5 whys."

When solving a problem, never make too many changes at once. It is best to make only one change at a time. Multiple fixes make it difficult to know which one solved the problem. Conversely, if the improvements make the problem worse, the more changes that were made at the same time, the harder it will be to identify which "improvement" caused it. It even may be necessary to reverse all the changes to get back to where you started.

Consider the problem-solving technique to be used. In the following list, I have a few favorites that I always use. They are the ones in italics.

- *Brainstorming*
- *"5 whys"*
- 5W + 1H
- Fishbone
- *Cause and effect with addition of cards (CEDAC)*
- Fault tree
- Six Sigma and the DMAIC process
- *The following are specific productivity tools, all of which I use: 5S, SMED, TPM, and RCM.*

Brainstorming

It is no longer politically correct to use this term because it apparently has medical connotations. Thought showering seems to be favored. So what is brainstorming? Have you ever sat in a bar and discussed a football game? "The ref never should have done that!" "Playing him was a mistake. Jimmy G would have been a much better choice." Each person at the table has his input, and then he jumps to a new topic to criticize. This is called *brainstorming*. The only difference between proper brainstorming and bar brainstorming is that in the real one, there are simple rules to follow, and the ideas are recorded. The points can be recorded directly onto a flip chart or onto Post-its. The main advantage of the bar technique would have to be the beer.

Some people prefer to record their ideas on Post-its and then put them on a chart or wall for analysis. This is not my preference. I like the team to sit around a table and then ask each person for one idea each, in rotation. One person should note the ideas on Post-its—one per idea. I believe that this form of interaction triggers new ideas, like telling jokes in a group, and the team ultimately ends up with more ideas.

The rules include the following:

- All ideas are good ideas.
- No criticism is allowed.
- One idea is suggested at a time.

▲ Aim for as many ideas as possible.
▲ Don't evaluate until ideas are exhausted.
▲ Record all ideas.
▲ "Yes but" is banned.

Try to get as many ideas as possible. The rules are designed for a purpose. Not following the guidelines either inhibits input through criticism or breaks the flow of ideas. Never try and solve the problems until all the ideas are exhausted. Once the stream of ideas starts, do all you can to ensure that the momentum keeps flowing.

The "5 Whys"

This is the technique learned by children, normally through experience. If you don't play a game with them—usually just after a day's work—they will ask "why" repeatedly until they get the root-cause answer, "I don't feel like it." It also can be used to solve serious issues such as the big bang theory! As a technique, it is wildly underrated.

Example 1

Q1: *Why* is there oil under the car?
A: The engine is leaking.

Q2: *Why* is the engine leaking?
A: The oil filter is leaking.

Q3: *Why* is the oil filter leaking?
A: The filter is not tight.

Q4: *Why* is the filter not tight?
A: The thread has worked loose.

Q5: *Why* has the thread worked loose?
A: It is the wrong filter for this car.

Q6: *Why* is it the wrong filter?
A: It is the wrong part number.

Q7: *Why* is it the wrong part number?
A: It was copied incorrectly.

Q8: *Why* has the number been copied incorrectly?
A: The manual is dirty, and the number is not clear.

Did you spot the deliberate mistake? There are eight whys, not five. We might only need to ask "why" three or four times. The number of times is not important. If we reach a person's name, we probably have gone too far. I recommend writing down the steps of the analysis in case the root cause does not work. This will help you to find out how the wrong path was followed. But what if there are more choices in the answers?

Table 7.1 Example 2

1	*Why* did the customer not buy the product?		
A	The salesperson did not persuade her to buy.		
2	*Why* did the salesperson not persuade the customer to buy?		
A	The salesperson was not good enough.	The customer did not want the product.	The customer wanted the product but did not buy.
3	*Why* was the salesperson not good enough?	*Why* did the customer not want the product?	*Why* did the customer not buy the product?
A	The salesperson has not been trained in sales.	The product was not what the customer needed.	The product was too expensive.
4	*Why* has the salesperson not been trained in sales?	*Why* was the product not what the customer needed?	*Why* was the product too expensive?
A	It was not considered necessary.	The salesperson misunderstood the need.	There was no suitable product.	The pricing policy sets the cost.
5	*Why* was training not considered necessary?	*Why* was the need not understood?	*Why* was there no suitable product?	
A	Sales are only a small part of the job.			

I could go on, but the result is not important. The method allows options sprouting out of the main column, like children on a family tree, when the answer is not clear-cut. In this case, some investigation or analysis would help. It is good practice, for example, to follow up on lost sales. Ask the customer why she did not buy.

Fishbone Diagram

The fishbone diagram once was my least favorite fault-finding tool until I discovered the cause and effect with addition of cards (CEDAC) diagram, which I will cover later. The fishbone diagram looks like the skeleton of a fish (Figure 7.3).

The main problem is entered in the nose. Ensure that it is well explained. It could include graphs or a photo if this helps with understanding of the issue.

The bones (or spines) originally had only the "4 M's." Once, all problems were reduced to one of the four: *man, machine, material,* or *method*. This is not an unrealistic assumption. Eventually, *measurement* was added to highlight how critical it is to have an understanding of the reliability, reproducibility, and accuracy of the measuring system. *Environment* was added to make people consider the location of a tool and the impact of its surroundings on the operation. (Imagine a stove sitting next to a freezer.)

The reason that I disliked the fishbone was that the bones get cluttered very quickly. A measurement issue might have three possible options (causes 7, 8, and 9), but each of these causes could have its own subcauses, and so on. Either you need to be able to write really small or use a blue whale as the fish. This leads us nicely to the CEDAC diagram.

CEDAC Diagram

Strictly speaking, my favorite is a *CEDAP*, where *P* stands for Post-its. But there is no such diagram.

Everything that holds true for the fishbone diagram is true for the CEDAC diagram. The main beauty of the CEDAC diagram is that when one spine gets cluttered, you can simply peel off the Post-its and use them to start to make a new fish, as in Figure 7.4. Using the earlier example to look more deeply into measurement issues, the *measurement* bone would

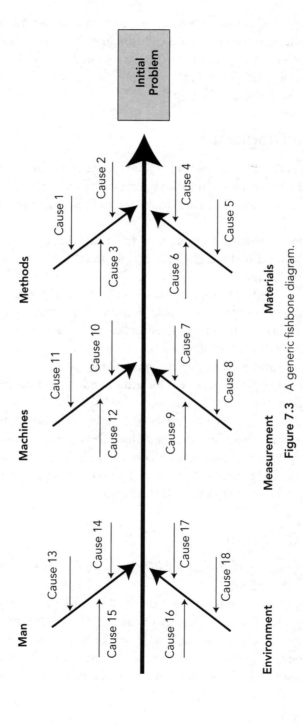

Figure 7.3 A generic fishbone diagram.

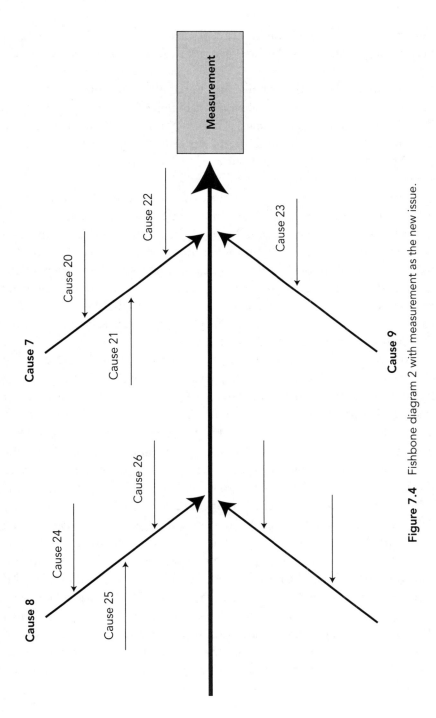

Figure 7.4 Fishbone diagram 2 with measurement as the new issue.

become the new "nose" and central spine, and causes 7, 8, and 9 would become the main bones. As the diagnosis develops, you could generate a whole family of fish.

The main bones do not need to be labeled with the "4 *M*'s." The problem "Car will not start" could have main bones labeled "Fuel," "Battery," "Starting technique," "Spark plugs," "Starter motor," and "Other electronics." The "4 *M*'s" are used to name the bones only as a guide as to what the possible options might be when the specific issues are unknown. They are simply memory prompts.

I thought I might just mention the concept of cause and effect (the *CE* of CEDAC). If I carry out an action and something happens because of it, the initial action is the *cause*, and the consequence (the thing that happened because of it) is the *effect*. If I hit a window with a hammer, the effect is likely to be a broken window. For every cause, there is an effect. There might be more than one effect. If I threw the hammer at the window, I might break the glass, but I also could damage the frame or break several items inside the room. The effect follows the cause. The glass does not break before I throw the hammer despite what I tell the judge!

When fault finding, brainstorm—think about all the possibilities. What else could be the cause of the effects I am seeing? The effect is the symptom. It is not easy to think outside the obvious. Consider a precision furnace, for example. The furnace has a process gas flowing through it at a precise pressure. The pressure gauge is very accurate, but the process does not have a stable output. To improve reliability, every equipment function is considered: the quality of the gas, the quality of the material it acted on, the temperature of the furnace, the flow of the gas, the measurement systems, and the skill of the people running the experiment. The effect patterns could be reproduced by either increasing or decreasing the pressure. But the pressure was perfect! The gauge and flowmeter were changed, but the problem remained.

When nothing was found, the team started to look wider. Eventually, someone noticed that the results were better on a sunny day, and with a bit of analysis and testing, the team found that the results were bad when it was raining. The cause turned out to be the outside pressure changes: A sunny day equates to high pressure and rain to low pressure. The cure was an external pressure gauge located on the roof and wired into the equipment pressure measurement system to add a correction. If the consequences are serious, try to think of everything.

It is difficult to think the unacceptable. But just to demonstrate that the "5 whys" have no limit, let's consider something really big—the big bang. See Table 7.2 for an example.

Table 7.2 Example 3: The Big Bang Whys

Q1	Why did the big bang happen?	
A	A point in space exploded.	
Q2	Why did the point in space explode?	
A	It was a spontaneous explosion.	The mass of the point became unstable, too great to sustain itself.
Q3	Why did the point explode spontaneously?	Why did the mass of the point become too great to sustain itself?
A	An unknown reaction occurred.	Gravity pulled in too much material.
Q4	Why did an unknown reaction occur?	Why did gravity pull in too much material?
A	Something in the point source must have changed.	It was the strongest source of gravity around.
Q5	Why did the point source change?	Why was it the strongest source of gravity around?
A	It became unstable.	It absorbed all the other masses within reach.
Q6	Why did it become unstable?	Why did it absorb all the other masses around it?
A	Something happened to it that caused it to become unstable.	Because of its gravity, it pulled the others into it.
Q7	Why did something happen to it?	Why did it not pull in more masses?
A	It had the capability of interacting with the space around it.	There became more mass than the gravity could sustain.
Q8	Why did it have the capability to interact with the space around it?	Why could it sustain no more mass?
A	It had a solid mass.	More mass made it more unstable.
Q9	Why did it have a solid mass?	Why did more mass make it unstable?
A	It was created by the effect of gravity.	An unknown reaction, possibly similar to plutonium's reaction to mass, caused a massive release in energy.

The answers are guesses—naturally. There should be multiple columns, each with a range of possible causes. I believe that the big bang did happen because all the galaxies are moving away from a point. But was it a point or a large volume in space? The cause must be something to do with the size. If you get small black holes and *supermassive* (huge) black holes, this suggests that they just keep growing, like Hoover bags, filling up through the action of all the black holes merging together until a point arises where they become one single, ultra-supermassive, possibly unstable mass. I associate what is happening with the way plutonium explodes when there is too much of it in one lump, but what if one black hole met an antimatter black hole?

If you can cause a micro–big bang in an accelerator and it is assumed that the original material (in the big bang) came from nothing, why would this mechanism limit itself to being a miniature big bang?

I also believe that there have been multiple big bangs throughout space, and our universe is only one of many, like one floating balloon among billions. If this is the case, we may find some galaxies moving in a different direction from the ones that came from our big bang because they came from a different big bang "point".

What about the galaxies speeding up and not being pulled back toward the center of the universe? Could it be that they are being attracted by the gravity of the external galaxies?

Have a go with the "5 whys."

In normal problem solving, not everyone will be happy with the causes identified by the team. So be tactful, and never make it personal. If you think that something may be possible, say so. You may be right, or you might trigger the right solution from someone else.

Fault-Tree Analysis

I used this technique before I knew that it had a name (Figure 7.5). It emulated the way I think. When I went to fix a problem, I would ask, "If I wanted to cause this fault or create these symptoms, how would I do it?" My original example was a tree for "sleeping in." I considered everything I could think of. However, when I used it for the first time with a class, I discovered that I had missed a whole range of issues. My own experience had huge holes in it! I never thought of insomnia, shifts, children, and late-night emergencies. To create a better tree, I should have used a group of

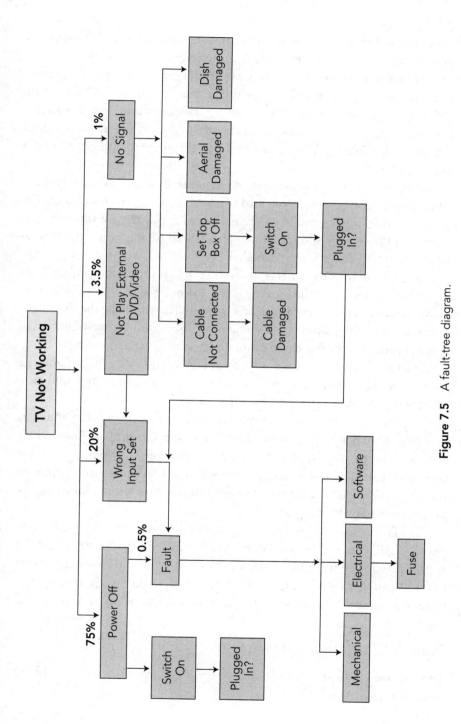

Figure 7.5 A fault-tree diagram.

people and brainstorming techniques. I assembled the following TV example by myself, so there will be holes in it, too.

If I had created the TV fault tree as part of a team, there would be more options, but I only want to illustrate the technique (Figure 7.5). To explain the diagram, I could think of three main reasons for the TV not working: the power being off, the wrong input channel being set, and no input signal. I considered a fourth that is also a check to eliminate the others: Would it play a DVD?

The power being off is a simple check, naturally followed by two others, which, if they don't work, would lead you to consider an electrical fault. The wrong channel input would be a greater possibility if the TV was switched "On" and there was still no picture. This would be the same situation as would be seen if there was no signal.

One advantage of fault-tree analysis is that you can add probabilities to the causes. The most likely cause is the power being off. I estimated it as being the case 75 percent of the time. Naturally, this would be the first check to make. The wrong input is becoming more likely with the introduction of home theater setups. You could estimate around 20 percent owing to them. The least likely option is no signal. Suppliers tend to be reliable, and the interconnections are hardwired and unlikely to work lose. You would have to set this probability low—maybe 1 percent?

The fault tree looks like a family tree, with new causes being added as experience increases. The same holds true for the probabilities. Their accuracy can be improved as the system is developed. When the chart is complete, you can start to compare it with the real causes to confirm or remove options. If you follow the probabilities, you should come to the proper solution with the least effort.

The fault tree is almost a visual representation of an *out-of-control action plan* (OCAP). The OCAP is a shopping list of the most likely causes with the appropriate checks included. Working down the list should be the fastest way to find a problem. Once more, if you find that the solution always seems to be far down the list, the order of the list is wrong and needs to be reset.

Step 7: Circulate the Results

The last two stages are simple but essential. They would be included in the control stage of DMAIC.

Improvements must be understood and practiced by everyone. The team also deserves recognition for its hard work. Circulating the results and displaying them on notice boards is a good way to achieve this. You want to develop a culture of improvement and get everyone involved. How better to do this than by giving others a chance to review the process and see how it was done. This also serves the function of a final check on the work—to highlight any issues before the system is made permanent.

Step 8: Embed the Solution

Training on new procedures no doubt will be required, but how much will depend on the complexity of the changes made. New operating instructions or procedures most likely will need to be written. At the very least, changes will need to be displayed as notices where the work is carried out. It is important that the team pass on any new procedures or techniques learned. Team members can lead or at least participate in new teams and pass on their skills.

The final stage is embedding the system into daily routines. Without embedding, the new system will not last. Responsibility for implementing the outcomes can be delegated to production operators, including recording the before and after implementation data, such as SPC. It also may be necessary to formalize any new maintenance routines or a program of quality checks.

Failure to embed the new process will give the impression that the improvements are an extra to the normal job. As such, the improvements probably will disappear at the first opportunity, which is normally when there is a run on production or when a breakdown costs time that needs to be made up. If the new system is not carried out, even just a couple of times, it is very likely that it will be lost.

What Are the Consequences of a Wrong Solution to a Problem Being Applied?

This is a difficult topic. There is always a chance that the wrong decision will be made. I remember reading an article many moons ago that claimed that more people died in hospitals owing to misdiagnosis than to any other cause. I wonder how true this is. I like to watch a TV show called *House* in which the doctor (House) tackles complex problems. He uses a range of analysis techniques, including brainstorming, to get a list of possible causes and often has to reduce the possible options by elimination, trying the most likely treatment to see if it works. You probably can guess that in a 45-minute show the first few ideas never work! Often, for dramatic effect, the patient nearly dies a few times. Mind you, on a piece of equipment, I have often seen random exchanges or improvements causing equipment death.

If you have a decision to make, what do you do? I am sure that we all logically consider the evidence, rationalize the costs, and pick the best option—*not*! We make decisions based on a number of factors: What will my wife think? Will this upset the person? Will my boss agree with this choice? Should I go for the obvious to avoid making the wrong choice? You can still be flexible, but maybe you should start by finding out what the real options are.

If you have a problem to solve. To solve it, you need information. If you already have access to the information, you can decide. If you do not have the information, you need to get it from other sources—colleagues, manuals, procedures, operators, and equipment suppliers. Colin Powell believes that we need at least 70 percent of the information before being able to make a reasonable choice. He believes that waiting for the 100 percent level takes too long and might affect the outcome, where time to solution or time to market is an issue (see Figure 7.6).

When you have made the choice, though, there is one more thing to consider before implementing it. What are the possible consequences if the choice is wrong? If the car has been breaking down a lot, should you use it to drive to the country for a picnic? What if it fails again? Can you get help? What if you can't? In a business environment, the wrong decision can cost money. The need is to avoid making the wrong choice or to limit the damage if you do. Thus you make a decision. The first question you need to ask is, "Is there any risk?"

If you decide that there is no risk, you can proceed to implement the decision. If you think that there is risk, you need to identify what the risks are and quantify them. If you don't know if there is a risk, you should assume that there is.

In solving a problem, there are usually a few alternative solutions, some of which will be better than others. You would brainstorm for the options and use the preceding decision techniques—perhaps try a fault-tree analysis

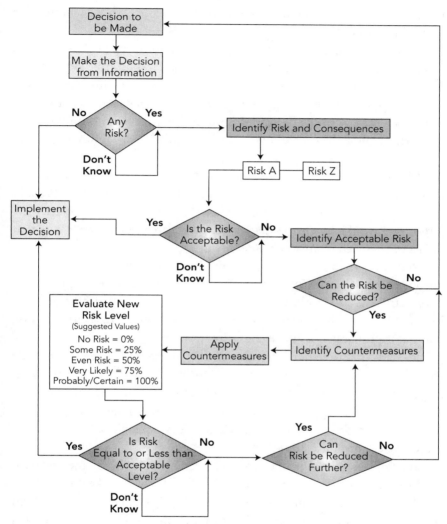

Figure 7.6 The decision flowchart.

or CEDAC diagram to see if you can predict and quantify what can go wrong. Even simple issues can have a detrimental effect.

Let's consider another problem. We are an operator down and will miss our production targets. The usual response is not to recognize the fact that it is a potential problem. We can just use another operator. Jamie used to work on the tool, and we have the written procedure, so we will just use him. What can possibly go wrong?

Worst case, Jamie can damage the tool or scrap a whole shift's production. What if he used the tool recently, say, last month? He probably will have no problems. What if it was last year, though? This is a different situation. But we still have our set of instructions. What if they are too hard to read or easy to misunderstand? So we show him how to do it. Is one explanation enough? It takes the average person 17 to 19 times on average to learn how to do a complex task properly.

We now find out that there are risks. We also have an idea about what they could cost us. We must decide whether the risk is acceptable. It could be a bit high, so what can we do? We need to reduce the risk—exactly the same as we would do with a safety issue. What can we do to make sure that the risk is reduced to an acceptable level?

Returning to the operator problem: What if we had someone check the output every half hour until we were sure that there were no problems? Better still, we also could confirm that the operator was not having any issues. What if the instructions were fail-safe? (We used to say idiot-proof, but that is probably not allowed any more.)

It seems that there are a few things that you can do to reduce the risk. If not, then you should decide whether the potential loss is acceptable. It might be that if the order does not go out tomorrow, you lose the sale. This would raise the acceptable level of risk. On the other hand, it might mean that you bring someone in on overtime or use two people and try all of the preceding countermeasures.

You will get to a point where you have reduced the risk on the first pass, tried to reduce the risk even more, but still find that the likelihood of the problem is too high. At this point, you need to go back to the initial problem and either look for another solution or refine the problem to suit what is possible. At the end of the day, you make the decision based on the best information and options you have available. If you have no information or do not decide at all, then the outcome is deserved.

In the example of the operator, the wrong choice will cost a night's production, which could be thousands of dollars. Senior management has to design a protocol of what is acceptable. In RCM, it is recommended in the specification that when any process parameter falls outside its tolerance, production has to stop. This is the spec. In life, the written spec is frequently ignored, and extra allowance is added to enable production to run. If this is the situation, then the written spec should be changed to the acceptable values. There is no point in having a specification if it is not intended to be used. The limits for the specification should be set while calmly sitting in a meeting and there are no problems to add undue influence. Otherwise, it is like shopping when you are hungry: You buy lots of extra food and junk. So the line manager should have guidance as to what is an acceptable loss. When I am faced with such a choice, I wonder what I would do if it was my own money I was risking.

Let's consider another example. I need some chemicals for a process, but I have none of the correct materials in store. A wider search was made, and Jimmy found a 10-year-old bottle in the back of a cupboard. The decision is, "Should I use the old chemicals?" If I say "Yes," is there any risk that something might go wrong? If not, then I can proceed with relative safety. If not . . . !

What are the risks?

1. The chemical may not be what it says on the label.
2. The chemical may be contaminated.
3. There could be some effect owing to the age, similar to that with medicines.

In use, the chemical is heated to a temperature at which it becomes a gas. It becomes a gas at a very specific temperature, just as ice melts or water boils at a specific temperature. It is the gas that we need. We must decide if the risk is acceptable? If contaminated, the lost product could cost $100,000, which is not an insignificant loss. The risk of the loss has to be reduced. Fortunately, as it turns out, the machine planned for production works in such a way that it can purify the chemical. But there still could be some unforeseen contamination.

I do not know what the engineer used to decide. I only know that he did choose to use the chemical. When he ran the machine, the chemical did *not* run as expected. Indeed, it became a gas at half the expected temperature. There also was another, less obvious variation around the weight of the material. Both issues were overlooked by the engineer and the operator. The machine ran the product for three days, 24 hours a day. It turned out that over the 10 years, the chemical had oxidized (rusted), and for the whole time, the machine was running on the "rust" component and not the actual chemical.

I had not developed my decision chart back then, but I did have worries when I heard of the decision to use the chemical. Had some thought been given to countermeasures, a plan could have been made to ensure that the two wrong settings were not used. Indeed, the machine could have been operated in its automatic mode, where it would have refused to run the product. Possibly, this is why the unit was run in manual. (A third clue that something was wrong!)

Making decisions can be difficult. If the consequences are serious, have a quick brainstorm. However, at the very minimum, you need to ask the questions. In the first example about the inexperienced operator, a complete night's product was written off. It was that company which asked me to create a training course on decision making.

I did not have a course at the time, but I was writing a chapter on problem solving. I found it really hard to put the idea into words. The diagram came more easily, which eventually led to this chapter. I can confirm that the diagram worked and provided real help. The diagram can be applied to any decision.

The best advice I can offer is that you always should seek input from colleagues when the decisions are not simple. I would bet that your colleagues also would like to have the same option.

GLOSSARY

5S
A technique for helping to improve the layout of a workplace, laboratory, or factory.

"7 wastes"
Lean Manufacturing recognizes seven main wastes that all production suffers. They are transportation, inventory, movement, waiting, overprocessing, overproduction, and defects. When any process step suffers from one of these wastes, it will not be efficient. If you are analyzing a process step, look for the impact caused by the "7 wastes."

Availability
A key component of OEE. What percentage of the day/week/year is a production tool available to be used to make product? Anything that prevents the tool running product is a potential issue.

Batch size
How many units are made at one time.

Big picture map
A strategic mapping process at a high level for finding the big issues and seeing the interaction of departments and data with process, quality, and inventory data.

Blitz
Also know as a *Kaizen blitz*. This is a concentrated project, often over a few consecutive days. Problems are found and fixed. Issues can arise if no actions have been taken to ensure that the project sustains.

Bottleneck
A part of a process that restricts the flow of product through it. The capacity of the bottleneck (assuming 100 percent efficiency) sets the productivity rate. A bottleneck often can be recognized by a buildup of raw material in front of a workstation.

Brainstorming
Also know as *thought showering*. This is a team analysis technique that creates a list of possible causes and options when considering issues.

Brown paper maps
Owing to the size of maps, mapping is carried out on walls, normally on rolls of paper or sheets that can be taken down when not in use or when visitors come.

Capacity
How many units a workstation or production line should be able to create per unit time.

Capacity map
An evolution of the process map that considers OEE and TOC. It looks at process steps and equipment in terms of their efficiencies: what they produce versus what they should be able to produce. It is very helpful for finding bottlenecks and poorly operating areas.

Culture
The personality of a company. It has a major bearing on the success of an improvement project.

Cycle time
This is defined as the time taken from the end of one process to the end of the next process.

Defects
One of the "7 wastes." Defects and reworks comprise one of the largest of the "7 wastes." A defect is anything that is not made right the first time.

Embedding
Making an improvement or other process become a normal part of a daily routine.

Flow
One of the five principles of lean. How a product or process progresses from start to end.

Future state map
The current state map represents the way the process runs now. There is also a future state map. This is the map that represents how the process will look when the improvements have been made.

Gemba
Going to see the source of the problem as opposed to diagnosing issues from a desk in an office.

Inventory
This is one of the "7 wastes." Inventory is all stock in a factory, including raw materials, work-in-process, and finished goods.

Kanban
Kanban uses a physical signal to warn of the need to reorder materials. It can be broken tape on the top of a box, a red line on a shelf marking the height of parts, a card in a box reminding the owner of the minimum stock, and so on. The most common is the two-box system. There are two containers of the same parts, nose to tail in the stores. When the first box is empty, the empty box is the flag used to trigger the reordering.

Mapping
Mapping a process or a full operation is a way to develop a visual representation of the individual steps. From the map, you also can highlight the specific areas where issues occur. There are several techniques available. See also, the big picture map, the value-stream map (VSM), the process map, and the capacity map.

Minor stops
Small interruptions or breakdowns that are thought to be too small to bother about. Some happen so often that if the cost of the losses were to be summed, the total annualized cost can be very high.

Movement
This is one of the "7 wastes." It refers to workplace layout and ease of access to the tools and materials needed.

On time and in full (OTIF)
This is a very important measure. Customer orders are not correct unless they arrive on time on the original date and with all the items ordered. Changing the date is a failure. Not making the delivery date shows a failure in the systems.

Overall equipment efficiency (OEE)
A measure that considers how well a machine runs compared with how well it should run. It considers quality, availability, and performance. More controversially, and not intended to demean anyone, I think that with proper use, a similar concept to OEE can be applied to people.

Overprocessing
This is one of the "7 wastes." Having more process steps than are needed for the product to fulfill customer requirements.

Overproduction
This is one of the "7 wastes." Making more product than the customer wants to buy.

Parking lot
A chart or notice board where issues are recorded for future use. They can arise during an analysis but may not be immediately relevant. They will be important, though, and you do not want to forget them. To maintain the flow of the original analysis, the parking lot is used.

Perfection
One of the five principles of Lean. To strive for perfection is the key to continuous improvement. Perfection is a goal in many processes. TPM, for example, promotes "zero fails"—no equipment breakdowns—perfection.

Plan-do-check-act (PDCA)
A process loop for implementing a new solution or process change. First, you need a *plan*. Then you *implement* it. Next, you *check* that the plan is doing what you expect of it. If it does not do what you expect, what is it doing wrong? You need to take *action* to correct it. Then you *plan* how you will implement it and so on through the loop.

Process map
The most detailed map. It looks at processes in detail to find steps that have problems and the effects of them. It can include any department steps from sales to shipping and includes specific documentation.

Production logs
A book or program that monitors the progress of production. Its purpose is to find out what interrupts normal running. Enough detail has to be entered to find subissues such as waiting for operators or engineers.

Profit
Turnover less costs and losses.

Pull
One of the five principles of Lean. Pull is customer demand. The perfect goal is to make only what the customer wants. In this way, you avoid waste.

RCM
Reliability-centered maintenance (RCM) was developed by the aircraft industry to reduce failures and breakdowns. RCM analyzes the root causes of issues using a system similar to failure modes and effects analysis (FMEA). It also systematically reviews each function of a tool and considers what the consequences of failure would be. It can be used in design of new equipment and in the development of maintenance procedures.

Reliability-centered maintenance
See RCM.

Right the first time
More accurately called *not right the first time*. Any product or task should be made with an expectation that it will be made correctly. Poor manufacturing often needs additional quality checks and rework and scrap waste resources (including money) and must be avoided.

Scottish Enterprise (SE)
An agency that works with the Scottish government and provides business support in a range of areas.

Scottish Manufacturing Advisory Service (SMAS)
A department with Scottish Enterprise that provides manufacturing support to companies.

Short-interval control (SIC)
A simple process that promotes monitoring production and other tasks and enables any deviation from the expected behavior to be detected as quickly as possible, minimizing any losses.

Single-minute exchange of die
See SMED.

SMED
Single-minute exchange of die. A technique developed to reduce the time taken to change a production line from one process to another. The time it takes to carry out changeovers is often the reason for companies overproducing.

Supermarkets
Small storage areas located as close as possible to where the parts are used.

SWOT
A high-level strategy technique. Create a flip chart split into four segments. Top left should be strengths; top right is the opposite, weaknesses; bottom

left is opportunities; bottom right is threats. Simply brainstorm for ideas and use Post-its because they make sorting and grouping results much easier.

Takt time
How many units we need to make per unit time to meet the customer demand. This also will help us to evaluate how many operators we need to achieve this rate.

Theory of constraints
See TOC.

TOC
The theory of constraints. Operator steps and equipment speeds in production lines tend to have a throughput—a rate at which work passes through the workstation. If one step is slower than the others, it can become a bottleneck that slows down the rate product can be produced. It also considers equipment capacity, which is the maximum rate product can flow through a workstation. See also OEE and the capacity map.

Total productive maintenance
Also known as TPM. It was developed initially by the Japanese in the 1950s, but its evolution continues. It is a full-blown factory productivity system. Originally, it was based on equipment maintenance and production issues, but it was expanded to include administration. It is argued that the name should be total productive manufacturing.

TPM
See total productive maintenance.

Transporting
This is one of the "7 wastes." This relates to the distance between workstations, stores, etc.

Value
One of the five principles of Lean. Value is something that adds to a process or to a customer's satisfaction levels. In Lean, we talk of value-adding. We

also have the opposite term: non-value-adding. Non-value-adding tends be caused by one or more of the "7 wastes."

Value stream
One of the five principles of Lean. The stream is process flow in terms of value or no value.

Value- and capacity-stream map
A variation of the value-stream map that allows the efficiency of the step to be considered.

Value-stream map (VSM)
The VSM measures value. The analysis stage has only two states: "value" and "no value." If value equates to a 1, no value would be a 0. Value is good; no value is bad.

Waiting
This is one of the "7 wastes." Waiting for parts, operators, materials, instructions, equipment, and so on are all non-value-adding tasks.

INDEX

Italic page numbers reference figures.

4 M's, 106
5 whys, 155, 157–159
5S, 126
    workplace layout, 2
7 wastes, 2, 5, 45–47
    bottlenecks, 88–89
    defects, 141–142
    impact of on process maps, 96–104
    inventory, 136–137
    movement, 137–138
    overprocessing, 140–141
    overproduction, 138–139
    overview, 135
    transporting, 136
    waiting, 27, 138
    See also TIM WOOD acronym

Allan, Colin, 108
Analyzing the process, 37–39
Arkwright, Richard, 123
Assumptions, 105
Availability:
    and big picture maps, 72–76
    in OEE, 13

Bar charts, 118, *119*
Big picture maps, 32, 40, 55–57
    availability, 72–76

Big picture maps (*Cont.*):
    as a closed loop, 41
    cost of the problems, 82–83
    current state maps, 83
    data flow and data boxes, 70–71
    diagram, *60*, *62*
    employee input, 55
    future state maps, 83–85
    information flow, 67–70
    inventory and stocking points, 78
    and *Kanban*, 63–64
    lead times, 66, 77–78
    main issues at each stage, 81–82
    material flow, 61, *62*
    materials, 64–67
    materials resource planning (MRP), 64
    on time and in full deliveries (OTIF), 76–77
    overview, 49–50
    parking lots, 59–61
    vs. process maps, 52–53, 89
    production logs, 53–55
    quality checkpoints, 78–80
    rework and scrap loops, 81
    simple diagram of, *42*
    snapshot assessments, 49
    solutions, 85
    SWOT diagrams, 50–52

Big picture maps (*Cont.*):
  what you need to know to make a useful map, 57–59, 80–81
Bill of materials (BOM), 145
Blame culture, 36–37, 141
Blending techniques, 3
Blitz interventions, 9
BOM. *See* Bill of materials (BOM)
Bottlenecks, 88–89
Brainstorming, 30–32, 156–157
Brown paper maps, 40

Capacity flows, 126
Capacity maps, 19
  birth of, 108–110
  cabinet manufacturer example, 116–121
  diagram, *113*
  improving the capacity, 114–116
  overview, 105–106, 110–114
Cause and effect, 162
  See also CEDAC diagrams
CEDAC diagrams, 159–164
Changeover board, 68, *69*
Cherry-picking processes, 3
Constraints, 110
Consultants, 28–29
Continuous improvement, 125–126
Cost, 28
  evaluating the cost of the problem, 144–146
Cross-functional teams, 141
Culture, 14–16
  blame culture, 36–37, 141
Current state maps, 83
Customer audits, 25
Cycle time, 95–96

Data boxes, 70–71
Data flow, 70–71

Defects, 46, 102, 141–142
  See also 7 wastes
Define, measure, analyze, improve, and control. *See* DMAIC
Diagnosis, 3–4
Direct employees, 7
  See also Indirect employees
DMAIC, 147
  circulating the results, 166–167
  creating the improvement team, 149–150
  defining the problem exactly, 150
  embedding the solution, 167
  flow chart, *148*
  identifying the problem to be solved, 147–149
  planning the project, 154–155
  quantifying the problem, 150–154
  solving the problem, 155–166
Duration of the project, 29

Eighth waste, 2, 146
Employee input, 55
Employee involvement, charting, 11
Enterprise resource planning (ERP), 54
ERP. *See* Enterprise resource planning (ERP)

Facilitators, 90–92
Family-run companies, 14–15
Fault-tree analysis, 164–166
Firefighting, 33
Fishbone diagrams, 159, *160*, *161*
Flow, 131–133
  in Lean Manufacturing, 27
  material flow, 61, *62*
Ford, Henry, 123
Forecasts, 100

Formulas:
  for OEE, 13
  for profit, 1, 11
Future state maps, 83–85

Gemba, 23
Glossary, 173–180
The Goal (Goldratt), 110
Goldratt, Eli, 110
Government agencies, 29

Hale, Dave, 43
Hines, Peter, 9
Histograms, 153

Improvement, continuous, 125–126
Indirect employees, 7
  See also Direct employees
Information flow, 67–70, *71*
Inventory, 45–46, 78, 97, 136–137
  See also 7 wastes

Japan Institute of Plant Maintenance (JIPM), 106
JIPM. See Japan Institute of Plant Maintenance (JIPM)

Kaizen, 147
  circulating the results, 166–167
  creating the improvement team, 149–150
  defining the problem exactly, 150
  embedding the solution, 167
  flow chart, *148*
  identifying the problem to be solved, 147–149
  planning the project, 154–155
  quantifying the problem, 150–154
  solving the problem, 155–166

Kanban, 97
  and big picture maps, 63–64

Large companies, 15–16
Lead time, 66, 77–78
Lean Manufacturing, 19
  and continuous improvement, 125–126
  five principles of, 26–27
  focus on the customer, 126–127
  Toyota Production System, 139
Logs, 53–55, 151
Losses, quantifying, 43–44

Magic wand concept, 38–39
Manchester Manufacturing Advisory Service, 135
Manufacturing advisory services (MAS), 29
Mapping, 19, 29–30
  general guidance, 32–35
  See also Big picture maps; Brown paper maps; Capacity maps; Process maps; Value- and capacity-stream maps; Value-stream maps (VSM)
MAS. See Manufacturing advisory services (MAS)
Mass production, 123–125
Material flow, 61, 62, 71
Materials, 64–67
Materials resource planning (MRP):
  and logs, 54
  and stock checks, 64
Maximizing utilization, 100
Medium-sized companies, 15
Minitab, 155
Minor stops, 24
Mobile phones, 128

Movement, 46, 97–98, 137–138
  See also 7 wastes
MRP. See Materials resource planning (MRP)

"Nail Game," 30–31

OCAP. See Out-of-control action plan (OCAP)
OEE. See Overall equipment efficiency (OEE)
On time and in full deliveries (OTIF), 76–77
Operational amplifiers, 108–110
Opportunities, 51–52
OTIF. See On time and in full deliveries (OTIF)
Out-of-control action plan (OCAP), 166
Overall equipment efficiency (OEE), 2–3, 12–14, 20, 126
  formula for, 13
Overprocessing, 46, 98–99, 140–141
  See also 7 wastes
Overproduction, 46, 99–102, 138–139
  See also 7 wastes

Parking lots, 59–61
Perfection, 134
  in Lean Manufacturing, 27
Performance:
  causes for poor production performance, 106
  in OEE, 13
Plan-do-check-act cycle, 35–37
Pollock, Agnes, 16, 38, 56
Post-its, 40–41, 92, 93, 112
Problem solving:
  5 whys, 155, 157–159
  brainstorming, 30–32, 156–157

Problem solving (*Cont.*):
  CEDAC diagrams, 159–164
  consequences of a wrong solution being applied, 167–172
  evaluating the cost of the problem, 144–146
  fault-tree analysis, 164–166
  fishbone diagrams, 159, *160*, *161*
  See also DMAIC; *Kaizen*; Root cause
Process maps:
  vs. big picture maps, 52–53, 89
  bottlenecks, 88–89
  creating the map, 92–96
  diagram, *94*
  impact of the 7 wastes on the first eight steps, 96–104
  introduction to, 40–43
  mapping a restaurant process, 89–90
  overview, 87–88
  simple diagram of with one main process and two customer options, *44*
  summary, 104–105
  the team, 90–92
  types of process layouts, *103*
Production. See Mass production; Overproduction
Production leveling, 100
Production logs, 53–55, 151
Production villages, 124
Profit, 1–2
  formula for, 1, 11
Projects:
  duration of the project, 29
  resistance to first projects, 4–9
  starting, 25–26
  time to do, 9–11

Pull, 100, 133–134, 139
  in Lean Manufacturing, 27
Push, 100

Quality, 81
  in OEE, 13
Quality checkpoints, 78–80
Quality checks, 141
Quantifying losses, 43–44

RCM. *See* Reliability-centered
  maintenance (RCM)
Reliability-centered maintenance
  (RCM), 33, 126
Resistance, 112
  to first projects, 4–9
Reworks, 81
Right first time, 22–23
Root cause, 23–24, 106, 143–144

Scottish Manufacturing Advisory
  Service (SMAS), 76, 107
Scrap loops, 81
Short interval control (SIC), 11–12, 80
SIC. *See* Short interval control (SIC)
Single-minute exchange of die
  (SMED), 2, 126
Slater, Samuel, 124
SMAS. *See* Scottish Manufacturing
  Advisory Service (SMAS)
SMED. *See* Single-minute exchange
  of die (SMED)
Snapshot assessments, 49
Solutions, 167–172
SPC. *See* Statistical process control
  (SPC)
Spreadsheets, 118
Springfield Rifle Company, 124
Starting projects, 25–26
Statistical process control (SPC), 155

Stock checks, 64
Stocking points, 78
Strengths, 50–51
Sustainability, 16–17
SWOT diagrams, 50–52

Takt time, 133
Targets, 16
Team leaders, 90–92
Teams, 90–92
  creating the improvement team,
    149–150
  cross-functional teams, 141
Techniques:
  adapting to new techniques, 2–3
  blending techniques, 3
Terminology, 173–180
Threats, 52
TIM WOOD acronym, 45–46, 96,
  135
  See also 7 wastes
Time savings, 7
Total productive maintenance
  (TPM), 2, 3, 126
Toyota Production System, 139
  See also Lean Manufacturing
TPM. *See* Total productive
  maintenance (TPM)
Training, 28
Transporting, 45, 96–97, 136
  See also 7 wastes
Tray exchange example, 20–22
Turnover, 1–2
Two-box *Kanban* system, 63–64, 97

University departments, 29
Unloading process, 107–108, *109*
Untapped human potential, 2, 146
  See also Wastes
Utilization, maximizing, 100

Value, 127–129
    adding some value, 20–22
    in Lean Manufacturing, 26
Value- and capacity-stream maps, 19
Value stream, in Lean Manufacturing, 26–27
Value-stream maps (VSM), 19, 129–131
Value-add plot, 83–85
Visual tracking, 141

VSM. *See* Value-stream maps (VSM)

Waiting, 27, 46, 96, 98, 116, 138
    See also 7 wastes
Wastes, 2
    See also 7 wastes
Weaknesses, 51
White board, 141
WIP. *See* Work-in-progress (WIP)
Work-in-progress (WIP), 78
Wrong solutions, 167–172